Early Praise for *Genetic Algorithms and Machine Learning for Programmers*

I really like the book; it's pitched nicely at the interested beginner and doesn't make undue assumptions about background knowledge.

➤ **Burkhard Kloss**
Director, Applied Numerical Research Labs

A unique take on the subject and should very much appeal to programmers looking to get started with various machine learning techniques.

➤ **Christopher L. Simons**
Senior Lecturer, University of the West of England, Bristol, UK

Turtles, paper bags, escape, AI, fun whilst learning: it's turtles all the way out.

➤ **Russel Winder**
Retired Consultant, Self-Employed

This book lifts the veil on the complexity and magic of machine learning techniques for ordinary programmers. Simple examples and interactive programs really show you not just how these algorithms work, but bring real-world problems to life.

➤ **Steve Love**
Programmer, Freelance

Genetic Algorithms and Machine Learning for Programmers

Create AI Models and Evolve Solutions

Frances Buontempo

The Pragmatic Bookshelf

Raleigh, North Carolina

Many of the designations used by manufacturers and sellers to distinguish their products are claimed as trademarks. Where those designations appear in this book, and The Pragmatic Programmers, LLC was aware of a trademark claim, the designations have been printed in initial capital letters or in all capitals. The Pragmatic Starter Kit, The Pragmatic Programmer, Pragmatic Programming, Pragmatic Bookshelf, PragProg and the linking *g* device are trademarks of The Pragmatic Programmers, LLC.

Every precaution was taken in the preparation of this book. However, the publisher assumes no responsibility for errors or omissions, or for damages that may result from the use of information (including program listings) contained herein.

Our Pragmatic books, screencasts, and audio books can help you and your team create better software and have more fun. Visit us at *https://pragprog.com*.

The team that produced this book includes:

Publisher: Andy Hunt
VP of Operations: Janet Furlow
Managing Editor: Susan Conant
Development Editor: Tammy Coron
Copy Editor: Jasmine Kwityn
Indexing: Potomac Indexing, LLC
Layout: Gilson Graphics

For sales, volume licensing, and support, please contact *support@pragprog.com*.

For international rights, please contact *rights@pragprog.com*.

ISBN-13: 978-1-68050-620-4
Book version: P1.0—January 2019

Contents

Preface

Have you ever heard the phrase "Coding your way out of a paper bag"? In this book, you'll do exactly that. In each chapter, you'll examine different machine learning techniques that you can use to programmatically get particles, ants, bees, and even turtles out of a paper bag. While the metaphor itself may be silly, it's a great way to demonstrate how algorithms find solutions over time.

Who Is This Book For?

If you're a beginner to intermediate programmer keen to understand machine learning, this book is for you. Inside its pages, you'll create genetic algorithms, nature-inspired swarms, Monte Carlo simulations, cellular automata, and clusters. You'll also learn how to test your code as you dive into even more advanced topics.

Experts in machine learning may still enjoy the "programming out of a paper bag" metaphor, though they are unlikely to learn new things.

What's in This Book?

In this book, you will:

- Use heuristics and design fitness functions
- Build genetic algorithms
- Make nature-inspired swarms with ants, bees, and particles
- Create Monte Carlo simulations
- Investigate cellular automata
- Find minima and maxima using hill climbing and simulated annealing
- Try selection methods, including tournament and roulette wheels
- Learn about heuristics, fitness functions, metrics, and clusters

You'll also test your code, get inspired to solve new problems, and work through scenarios to code your way out of a paper bag—an important skill for any competent programmer. Beyond that, you'll see how the algorithms explore problems, and learn, by creating visualizations of each problem. Let this book inspire you to design your own machine learning projects.

Online Resources

The code for this book is available on the book's main page[1] at the Pragmatic Bookshelf website. For brevity, the listings in this book do not always spell out in full all the include or import statements, but the code on the website is complete.

The code throughout this book uses C++ (>= C++11), Python (2.x or 3.x), and JavaScript (using the HTML5 canvas). It also uses matplotlib and some open source libraries, including SFML, Catch, and Cosmic-Ray. These plotting and testing libraries are not required but their use will give you a fuller experience. Armed with just a text editor and compiler/interpreter for your language of choice, you can still code along from the general algorithm descriptions.

Acknowledgments

I would like to thank Kevlin Henney, Pete Goodliffe, and Jaroslaw Baranowski for encouraging me as I started thinking about this book. Furthermore, I would like to thank the technical reviewers, Steve Love, Ian Sheret, Richard Harris, Burkhard Kloss, Seb Rose, Chris Simons, and Russel Winder, who gave up lots of spare time to point out errors and omissions in early drafts. Any remaining mistakes are my own.

Frances Buontempo

---

1. https://pragprog.com/book/fbmach/genetic-algorithms-and-machine-learning-for-programmers

CHAPTER 1

Escape! Code Your Way Out of a Paper Bag

This book is a journey into *artificial intelligence* (AI), *machine intelligence*, and *machine learning* aimed at reasonably competent programmers who want to understand how some of these methods work. Throughout this book, you'll use different algorithms to create models, evolve solutions, and solve problems, all of which involve escaping (or finding a way into) a paper bag. Why a paper bag?

In a blog post, Jeff Atwood, co-founder of Stack Overflow, reflects on many programmers' inability to program.[1] He quotes various people saying things like, "We're tired of talking to candidates who can't program their way out of a paper bag."

With that in mind, the paper bag escapology is a perfect metaphor and makes a great case study for applying the various algorithms you'll learn. Plus, this is your chance to stand out from the pack and break out of the proverbial bag.

The problems presented throughout this book demonstrate AI, machine learning, and statistical techniques. Although there's some overlap between the three, most will stick with machine learning. However, it's important to understand that all of them share a common theme: that a computer can learn without being explicitly programmed to do so.

AI isn't new. John McCarthy, the inventor of the Lisp programming language, coined the term artificial intelligence in a proposal for a conference in 1956. He proposed an investigation, writing:

> The study is to proceed on the basis of the conjecture that every aspect of learning or any other feature of intelligence can in principle be so precisely described that a machine can be made to simulate it. An attempt will be made to find how to

---

1. blog.codinghorror.com/why-cant-programmers-program

make machines use language, form abstractions and concepts, solve kinds of problems now reserved for humans, and improve themselves.[2]

Recently, the topic of AI has surfaced again. This is likely because of the increase in computing power.

With today's modern personal computer, AI is more accessible. Many companies now offer automated chatbots to help us online. Robots explore places that are far too dangerous for humans. And thanks to the many programming libraries and frameworks available to handle the complicated mathematics, it's possible to find a neural network implementation, train it, and have it ready to make predictions within minutes. In the 1990s, you'd have to code this yourself, and then wait overnight while it chugged through data.

Many examples of AI involve computers playing games like chess, Breakout, and Go.[3] More generally, AI algorithms solve problems and explore data looking for patterns. The problem-solving part of AI is sometimes called machine learning—which includes analyzing data, allowing companies to spot trends and make money.

Machine learning is also an old term. Arthur Samuel, who built the first self-learning program that played checkers or draughts, introduced the term in 1959.[4] He researched ways to make programs get better at playing games, thereby finding general-purpose ways to solve problems, hence the term *machine learning.*

Machine learning has become a buzzword recently. It's a huge topic, so don't expect to master it any time soon. However, you can understand the basics if you start with some common ideas. You might even spot people trying to blind you with science and think of probing questions to ask:

- How did you build it? If it needs data to learn, remember: Garbage in, garbage out. Bias in, bias out.[5]

- How did you test it? Is it doing what you expect?

- Does it work? Have you got a solution to your problem?

- What parameters did you use? Are these good enough or will something else work better?

---

2. aaai.org/ojs/index.php/aimagazine/article/view/1904
3. https://www.wired.com/story/vicarious-schema-networks-artificial-intelligence-atari-demo/
4. en.wikipedia.org/wiki/Arthur_Samuel
5. www.designnews.com/content/bias-bias-out-how-ai-can-become-racist/176888957257555

- Does it apply generally? Or does it only work for your current problem and data?

Let's Begin

You'll start your journey by plotting points that are connected by lines. This is not a formal machine learning algorithm, but it introduces a few important terms and provides a clearer picture of what machine learning is and why it matters. Later, you'll use a decision tree and launch into a more formal machine learning algorithm.

The programming language used in this exercise is Python, although the language itself isn't important. In fact, throughout this book, you'll use a combination of Python, C++, and JavaScript. However, you can use any language you want. Some people claim you need to use general-purpose computing on graphics processing units (GPGPU), C++, Java, FORTRAN, or Python to implement AI algorithms. For certain applications, you may need a specific tech stack and a room full of powerful server machines, especially if you're looking to get power and speed for *big data* applications. But the truth is, you can implement any algorithm in the language of your choice; but keep in mind, some languages run quicker than others.

Get Out of a Paper Bag

For this exercise, imagine there's a paper bag with a turtle inside. The turtle is located at a specific point, and his task is to move to different points within his environment until he makes it out of the bag. He'll make a few attempts, and you'll guide his direction, telling him when to stop. To help see what's going on, you'll draw a line that joins the points together. You'll also keep these points around for reference in case the turtle wants to try them again later. By the way, there's nothing stopping the turtle from busting through the sides.

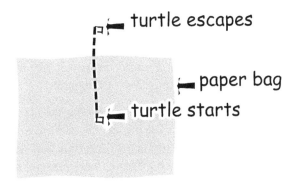

Guided by a *heuristic*, the turtle can make it out alive. A heuristic is a guiding principle, or best guess, at how to solve a problem. Each attempt made is considered a *candidate solution*. Sometimes those solutions work, and sometimes they fail. In the case of your wandering turtle, you need to be careful that he doesn't end up going around in circles and never escaping. To prevent that from happening, you need to decide on the *stopping criteria*. Stopping criteria is a way to make sure an algorithm comes out with an answer. You can decide to stop after a maximum number of tries or as soon as a candidate solution works. In this exercise, you'll try both options.

It's time to get into the mission.

Your Mission: Find a Way Out

To solve this problem, you have lots of decisions to make:

- How do you select the points?
- When do you stop?
- How will you draw the lines?

No matter how precise a description of an algorithm is, you always have choices to make. Many require several parameters to be chosen in advance. These are referred to as *hyperparameters*. Trying to tune these is a difficult problem, but each algorithm presented comes with suggested values that work. They may not be optimal, but you can experiment with these to see if you can solve the problems more quickly, or use less memory.

Remember, you need some kind of stopping criteria too. For this problem, you'll be trying two methods: guessing how many steps are needed, and letting the turtle move around until he escapes. For other problems, it's simpler to try a fixed number of iterations and see what happens. You can always stop the algorithms sooner if it solves the problem. Although, sometimes you might let them run past your first guess.

There are a few ways in which the turtle can escape the bag. He can start in the middle and move in the same direction, one step at a time, moving along a straight line. Once he's out, he'll stop, which means you don't need to build in a maximum number of attempts. You do, however, need to choose a step size—but beyond that, there's not much left to decide.

The turtle can also move forward a step and then change direction, repeatedly, increasing the step size each time. Taking an increasing step is a heuristic you can use to guide the turtle. Whichever direction you pick, the turtle is likely to end up outside the bag since he takes bigger steps each time. Using a fixed angle to change direction and linearly increasing steps will build a *spirangle*.[6] A spirangle is like a spiral, but it has straight edges. Therefore, with this type of movement, the turtle will leave a spirangle trail behind.

If the wandering turtle turns through a right angle, he'll build up a rectangular, or four-angle spirangle. Starting with a smaller step size, he moves forward and turns through 90 degrees, twice. He increases the step size and does this again—forward, turn, forward, turn. By starting at the small circle, he'll leave a trail like the one in the following figure:

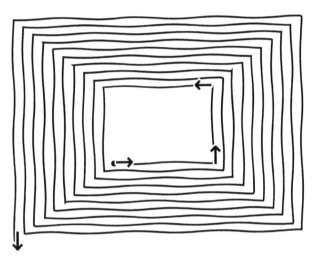

The arrows show which way he's moving. By choosing different angles, you get different shapes. If you can't decide what to try, pick a few different angles at random and vary at what point he changes the step size.

To recap, the turtle can move in straight lines or spirangles. He can also make lots of concentric shapes. For example, drawing a small square, then a larger one, and so on until he's drawn a few outside the bag. He'll have to jump to do this. But as long as he draws at least one point outside of the bag, he succeeds.

Of course, the turtle can also pick moves at random, but you'll have no guarantee that he'll end up on the outside of the bag. In fact, many of the algorithms in this book use randomness, whether they be random points in space or random solutions. However, these algorithms will either make candidate solutions guide each other, or they will compel their movement to behave in ways more likely to solve the problems. Learning needs more than random attempts, but it can start there.

---

6. en.wikipedia.org/wiki/Spirangle

How to Help the Turtle Escape

The turtle knows when to stop and has a few ways to pick the next points. We can pull these methods together into a program to try them all out. We want to be able to see what he's up to as well. The Python turtle package is ideal for showing movement from one point to another, and spirangles are often used to demonstrate its power. It comes with Python, so you don't need to install anything else. That's handy!

Turtle graphics pre-date Python, originating from the Logo programming language, invented by Seymore Papert.[7] The original version moved a robot turtle. He wrote a significant book with Marvin Minsky *Perceptrons: an introduction to computational geometry [MP69]* paving the way for later breakthroughs in AI, making the turtle package an excellent place to start discovering AI and machine learning.

Turtles and Paper Bags

When you import the package, you get a default, right-facing turtle with a starting position of (0, 0). You can choose your turtle shape, or even design your own. This turtle can rotate 90 degrees left, 90 degrees right, or any angle you need. He can also move forward, backward, or goto a specific location. With a little help, you can even get him to draw a paper bag, like this:

Escape/hello_turtle.py
```python
import turtle

def draw_bag():
  turtle.shape('turtle')
  turtle.pen(pencolor='brown', pensize=5)
  turtle.penup()
  turtle.goto(-35, 35)
  turtle.pendown()
  turtle.right(90)
  turtle.forward(70)
  turtle.left(90)
  turtle.forward(70)
  turtle.left(90)
  turtle.forward(70)

if __name__ == '__main__':
  turtle.setworldcoordinates(-70., -70., 70., 70.)
  draw_bag()
  turtle.mainloop()
```

---

7. https://en.wikipedia.org/wiki/Turtle_graphics

In the main function, on line 17, setworldcoordinates sets the window size. When you set your window size, be sure to pick something larger than the paper bag otherwise you won't see what the turtle is doing. Line 19, calls mainloop, which leaves the window open. Without the last line, the window shuts immediately after the turtle makes his move.

On line 4, you set the turtle's shape. Since the turtle starts at the origin, move him left and up on line 7. Because he starts off facing right, rotate him by 90 degrees, on line 9, so that he faces downwards. Then move him forward by 70 steps on line 10. Keep turning, then moving forward to outline the paper bag.

The finished bag is 70 units across, from x=-35 to +35, and 70 units high, also from y=-35 to +35. When you're done, you'll see the three edges of the bag and the turtle:

Now that you have a paper bag and know how to move a turtle, it's time to get to work.

Let's Save the Turtle

The goal is to help the turtle escape the bag you saw earlier on page 6. The easiest way is to make him move in a straight line. He might then march through the sides of the bag. You can constrain him to only escape through the top, but let him go where he wants for now. When he's out, you need to get him to stop. But how do you know when he's out? The left edge of the bag is at -35, and the right is at +35. The bottom and top are also at -35 and +35, respectively. This makes checking his escape attempts easy:

Escape/escape.py
```
def escaped(position):
    x = int(position[0])
    y = int(position[1])
    return x < -35 or x > 35 or y < -35 or y > 35
```

Now all you need to do is set him off and keep him going until he's out:

Escape/escape.py
```python
def draw_line():
    angle = 0
    step = 5
    t = turtle.Turtle()
    while not escaped(t.position()):
        t.left(angle)
        t.forward(step)
```

Simple, although a little boring. Let's try some concentric squares.

Squares

To escape using squares, the turtle will need to increase their size as he goes. As they get bigger, he'll get nearer to the edges of the paper bag, eventually going through it and surrounding it. To draw a square, move forward and turn through a right angle four times:

Escape/escape.py
```python
def draw_square(t, size):
    L = []
    for i in range(4):
        t.forward(size)
        t.left(90)
        store_position_data(L, t)
    return L
```

Store the position data, including whether or not it's in or out of the paper bag:

Escape/escape.py
```python
def store_position_data(L, t):
    position = t.position()
    L.append([position[0], position[1], escaped(position)])
```

You'll need to choose a number of squares to draw. How many do you think you need to get the turtle out of the bag? Experiment if you can't work it out. Now, move your turtle to the bottom left corner and draw a square, increasing the size as you go:

Escape/escape.py
```python
def draw_squares(number):
    t = turtle.Turtle()
    L = []
    for i in range(1, number + 1):
        t.penup()
        t.goto(-i, -i)
        t.pendown()
        L.extend(draw_square(t, i * 2))
    return L
```

You extend your list L of positions each time your turtle draws a square so you can save them:

Escape/escape.py

```
def draw_squares_until_escaped(n):
  t = turtle.Turtle()
  L = draw_squares(n)
  with open("data_square", "wb") as f:
    pickle.dump(L, f)
```

You'll use this data in the next chapter.

Spirangles

The turtle can also draw various spirangles by deciding an angle to turn through. If he turns through 120 degrees three times and keeps the step size the same, he'll draw a triangle. Increase the step forward each time, and he makes a spirangle with three angles:

Escape/escape.py

```
def draw_triangles(number):
  t = turtle.Turtle()
  for i in range(1, number):
    t.forward(i*10)
    t.right(120)
```

Try out other angles too. In fact, try something random:

Escape/escape.py

```
def draw_spirals_until_escaped():
  t = turtle.Turtle()
  t.penup()
  t.left(random.randint(0, 360))
  t.pendown()

  i = 0
  turn = 360/random.randint(1, 10)
  L = []
  store_position_data(L, t)
  while not escaped(t.position()):
    i += 1
    t.forward(i*5)
    t.right(turn)
    store_position_data(L, t)

  return L
```

Try this a few times, and save the points the turtle visits:

Escape/escape.py

```
def draw_random_spirangles():
  L = []
  for i in range (10):
    L.extend(draw_spirals_until_escaped())

  with open("data_rand", "wb") as f:
    pickle.dump(L, f)
```

Unlike the squares, you let the algorithm decide when to stop. You guessed in advance how many squares to draw on page 8 in order to have some fall outside the paper bag. This time, you baked some knowledge, or intelligence, into your algorithm. You'll discover various ways to do this in each chapter coming up.

Time to Escape

You can call any of these functions via main. Use the argparse library to check which function to call:

Escape/escape.py

```
if __name__ == '__main__':
  fns = {"line": draw_line,
      "squares": draw_squares_until_escaped,
      "triangles": draw_triangles,
      "spirangles" : draw_random_spirangles}

  parser = argparse.ArgumentParser()
  parser.add_argument("-f", "--function",
      choices = fns,
      help="One of " + ', '.join(fns.keys()))
  parser.add_argument("-n", "--number",
                default = 50,
                type=int, help="How many?")
  args = parser.parse_args()

  try:
    f = fns[args.function]
    turtle.setworldcoordinates(-70., -70., 70., 70.)
    draw_bag()
    turtle.hideturtle()
    if len(inspect.getargspec(f).args)==1:
      f(args.number)
    else:
      f()
    turtle.mainloop()
  except KeyError:
    parser.print_help()
```

You need to choose how many squares or triangles to draw, so you need to provide a number for these. The line and spirangles move until they're done. Your algorithm decides when to stop, so you don't have to. If you put all of your code in a file named escape.py, you can call it like this:

```
python escape.py --function=line
python escape.py --function=triangles --number=8
python escape.py --function=squares --number=40
python escape.py --function=spirangles
```

Did It Work?

Yes, you managed to code your way out of a paper bag in a number of different ways. Your first *deterministic* approach sent the turtle in a line, straight out of the bag:

Perhaps bursting out the side of the paper bag seems to be wrong. We will try other algorithms over the course of the book which avoid busting out of the sides. After the straight line, the turtle built squares that got larger and larger until some landed outside of the paper bag. If the turtle drew forty squares, spaced one unit apart, you see something like this:

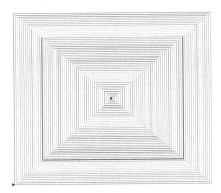

Finally, you made a variety of spirangles. If you picked 8 as the stopping criteria for the 120-degree turn, your turtle would have ended up outside the paper bag:

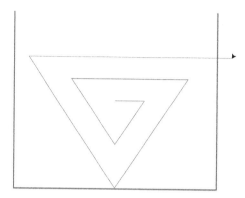

Rather than experimenting with different paths yourself, you can let the machine figure it out for you. All you need to do is give it a target to achieve, and let it go to work. When it's done, you will have several points outside of the bag:

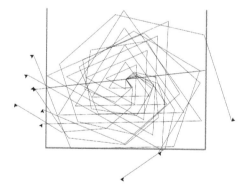

It might look a bit untidy, but you solved your first problem, and your algorithm includes some elements of machine learning: you wrote a function to decide if you had a viable solution and used this as its stopping criteria. You also used the heuristic of taking larger and larger steps. Throughout this book, you'll use *fitness* and *cost* functions to measure how good a solution is, and to compare solutions and pick better attempts.

You also tried several random variables and stopped when you succeeded. Many machine learning algorithms do this: they try a *stochastic search* (i.e., trying some random solutions). The idea of *learning* comes from building up better solutions iteratively.

Over to You

Turtle graphics exist for many programming languages, though unlike Python, most are implementations available as libraries. You can even build your own to use with your favorite language. You'll revisit turtles in the final chapter, Chapter 10, *Optimize! Find the Best*, on page 187. Until then, you'll use other ways to draw results.

If you wish to explore turtle graphics further, you may want to read about L-systems. The biologist Aristid Lindenmayer invented these to mathematically describe plants growing. Your spirangles grew iteratively. L-systems grow recursively, following simple looking rules made of symbols. Some symbols mean go forward, say F, or turn, maybe - for left and + for right. L-systems have rules built from these basic symbols, like:

```
X=X+YF+
Y=-FX-Y
```

You start this off with an *axiom*, like X, replacing the symbols as you encounter them. Since this uses recursion, you should keep track of the number of calls to stop somewhere. For example, this gives a curve known as a dragon:

```
Escape/dragon.py
from turtle import*

def X(n):
  if n>0:  L("X+YF+",n)
def Y(n):
  if n>0:  L("-FX-Y",n)

def L(s,n):
  for c in s:
    if   c=='-': lt(90)
    elif c=='+': rt(90)
    elif c=='X': X(n-1)
    elif c=='Y': Y(n-1)
    elif c=='F': fd(12)

if __name__ == '__main__':
  X(10)

  mainloop()
```

Like this:

Warning, it takes several minutes to render. If you search, you'll find ways to grow a variety of curves, including ferns and trees.

In the next chapter, you'll work through another algorithm using a divide and conquer approach (think sorting method). You can recursively split any data until it's in relatively pure groups or categories, such as inside or outside a paper bag. Ultimately, this creates a decision tree, indicating where it made different branches. It also predicts the category or group of new data not previously seen.

CHAPTER 2

Decide! Find the Paper Bag

In the previous chapter, you moved a turtle around and helped him escape a virtual paper bag. As he moved about, you drew lines, squares, and spirangles. You also saved the points he visited to a data file, noting if these points were located inside or outside of the bag.

In this chapter, you'll use the point data to build a *decision tree* and help the turtle find the paper bag. Because the data contains information about whether or not a point is inside the bag, you'll be able to *classify* sets of points. Once you have a decision tree, you'll turn it into a *ruleset* and *prune* it back to locate the paper bag.

A decision tree is a type of *classifier*. Many classifiers work like a black box—you feed in data, and it returns a prediction. Because a decision tree classifier is human-readable, you'll be able to determine why it gives its prediction and tweak it for better results. This *supervised learning algorithm* uses *training data* to find ways to predict the unseen data. The data has *features*, or x values, and a *category*, or target y value. Once trained, you can test your model on unseen data—if you're happy with the results, you can apply it to new data.

Decision trees can be used on all kinds of data. For example, they can split a list of chemicals into harmful or not harmful.[1] They can process credit card applications and assess them as low risk, medium risk, and high risk; they can even try to detect fraud.[2]

There are many ways to build a decision tree; in this chapter, you'll build one in Python using the Iterative Dichotomiser 3 method (ID3).[3] J.R. Quinlan invented ID3 algorithms in 1985, so they have a long history.

---

1. www.ncbi.nlm.nih.gov/pmc/articles/PMC2572623/
2. http://www.ijcsmc.com/docs/papers/April2015/V4I4201511.pdf
3. http://hunch.net/~coms-4771/quinlan.pdf

Building a decision tree will teach you one way to model data. You will see how to use *domain knowledge* and *entropy* to guide your tree's growth. Of course, you can use other methods—which you will do later in this book—but ID3 uses entropy as a *heuristic* so you will start there. A heuristic provides a best guess or shortcut to solve a problem and often comes in the form of *fitness*, *objective*, or *cost* functions. Each function tells you how well your algorithm is doing.

Your Mission: Learn from Data

Decision trees come in two forms: *classification trees* and *regression trees*. In both cases, you ask questions and take a branch depending on the answer. Once you reach a *leaf node*, you reach a decision. The questions asked can be *categorical* (e.g., which color) or *numerical* (e.g., how high). For a classification tree, the leaf is a category, like inside or outside a paper bag. For a regression tree, the leaf is an equation that gives a numeric value.

You can present your decision tree as a tree or a list of rules. In tree form, it looks like a flowchart. You can then transform this flowchart into a list of if-then-else rules by writing down the questions at each branch. You can also transform the rules into a tree. Each if statement makes a branch, and the then and else statements make sub-trees or leaves.

There are two main ways to build a decision tree: *bottom-up* and *top-down induction of decision trees*. The bottom-up approach builds a classifier from one data item at a time, whereas the top-down approach starts with all of the training data and then gradually divides it.

With the bottom-up approach, let's say you have a list of data that's classified as "good" or "bad" and includes letter and number features:

```
data = [['a', 0, 'good'], ['b', -1, 'bad'], ['a', 101, 'good']]
label = ['letter', 'number', 'class']
```

You can make a rule from the first data item using Pythonesque pseudocode.

```
if letter == 'a' and number == 0 then
  return 'good'
else
  return 'No idea'
```

You can then gradually relax existing rules or add new rules as more data is considered. When you use the next data item, you add another rule:

```
if letter == 'a' and number == 0 then
  return 'good'
```

```
else if letter == 'b' and number == -1 then
  return 'bad'
else
  return 'No idea'
```

The third data item lets you collapse these down. You already have letter 'a' mapping to 'good', and it doesn't matter what the number is:

```
if letter == 'a' then
  return 'good'
else
  return 'bad'
```

In contrast, the top-down approach uses all of the training data and gradually divides it up to build a tree. When the letter is 'a', you get 'good'; when you have 'b', you get 'bad'. You can encode this as a decision tree:

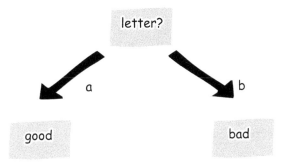

For this small dataset, the split point is easy to find. However, as you add more data, you'll need a way to determine on which feature to split. As mentioned earlier, you'll use entropy. Entropy has a formal definition in thermodynamics relating to the chaos of a system.[4] In *information theory*, it measures uncertainty. If you toss a two-headed coin, you can be certain you will get heads every time. If you toss a normal coin, you have a 50/50 chance of getting heads. The normal coin has more entropy. The coin type allows you to predict what can happen in the future. In the case of the lost turtle, you don't have heads or tails, but you do have x and y coordinates, and it works in much the same way.

Using a Python dictionary to represent the trees, you can start with an empty dictionary: tree={}. The key tells you which attribute to split on, and the value tells you what to do with split data. This'll either be a category or another tree.

For the letter and number data on page 16, you can make the letter a key which represents the split or branch. The corresponding value will then need

---

4.　en.wikipedia.org/wiki/Entropy

two leaf nodes (one for each value)—these are also dictionaries. One maps 'a' to 'good' and the other maps 'b' to 'bad'. Your tree looks like this:

```
tree = {'letter': {'a': 'good', 'b': 'bad'}}
```

Divide Your Data

Once you have a way to choose features for your branches, you can partition your data recursively in a similar way to quicksort. Think about how quicksort works:

- Pick a pivot; one element in the data.

- Rearrange the data into two groups; less than or equal to the pivot in one, everything else in the other.

- Apply the first two steps to each group, until you have groups of one or zero elements.

Quicksort uses a pivot point to divide the data into low and high values. A decision tree partitions the data using a feature instead of a pivot, but still recursively builds up a decision tree. By keeping track of the features on which you split, you can report the decision tree you built and try it on any data. You can also transform the tree into equivalent rules. This gives you a choice of ways to report what you discover.

You sometimes end up with lots of rules. In the worst case, you can get one rule per training data item. There are various ways to prune these back. Concerning the turtle point dataset, you can use the fact that the paper bag was square to transform a large ruleset into succinct rules.

How to Grow a Decision Tree

You build a tree from leaf nodes and sub-trees. The algorithm looks a lot like quicksort, partitioning the data and proceeding recursively:

```
ID3(data, features, tree = {}):
  if data is (mostly) in same category:
    return leaf_node(data)
  feature = pick_one(data, features)
  tree[feature]={}
  groups = partition(data, feature)
  for group in groups:
    tree[feature][group] = ID3(group, features)
  return tree
```

You partition the data into groups with the same value of your chosen feature. You build up sub-trees and make a leaf node when all of the data is in the same category—or it is mostly in the same category. This might be just one data item.

To decide a feature on which to partition the data, you can pick a feature at random, then build a *random forest*[5] and vote to form a decision. Unfortunately, there's not space to cover forests in this book but they're worth trying out. Instead, for this exercise, you'll build an ID3 decision tree using a more direct approach.

How to Decide the Best Feature

You can use all kinds of criteria for selecting features. Let's think generally first. Consider four points, (0, 0), (1, 0), (0, 1), and (1, 1). Suppose the first two are inside your bag and the last two are outside:

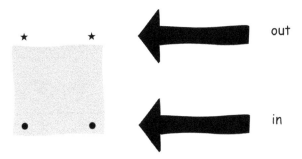

You only have two features from which to choose: the x and y values of the coordinates. The x coordinate can be inside or outside of the bag regardless of whether it's a value of 0 or 1. However, the y coordinate is only outside of the bag if its value is set to 1. With that knowledge, you can make this into a decision tree:

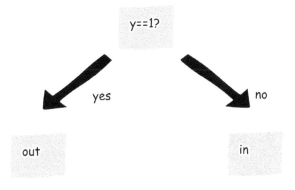

---

5. en.wikipedia.org/wiki/Random_forest

Of course, some points can be below the bottom of the bag or beyond its edges, so this isn't the full story. ID3 only knows what to do with feature values it was trained on. Only four possible points are using these x and y values, so you cannot use the decision tree it makes on any other data points. You will be able to make trees for other data using the same method though. You can try your algorithm on a larger paper bag, with edges at x=-1, x=1, y=-1, y=1, and use five training points: the center (inside the bag), and four points on the edge:

```
data = [[0, 0, False],
        [-1, 0, True],
        [1, 0, True],
        [0, -1, True],
        [0, 1, True]]
label = ['x', 'y', 'out']
```

You can now make new combinations of x and y values you haven't included in the training data, such as [1, 1]. You will be able to build a decision tree that classifies this new coordinate:

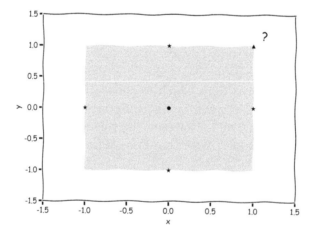

For now, let's consider how to use the four data points. You want to find the *purity* a feature gives you when you use it to partition your data. For the four points, using the y coordinate gave you two pure groups. There are many ways to measure this purity. Some use probabilities, and others use formal statistical measures. This advanced topic is not covered in this book, but a web search for "decision tree split quality" should get you started. However, since entropy doesn't need detailed statistical knowledge, it's a good place to start. You can try other ways to select features once you have learned how to build a decision tree.

Entropy tells you how *homogeneous* a dataset is. Entropy captures the concept of randomness or chaos in a system. Data with less entropy contains more structure and can be compressed more easily. If everything is identical, like with a two-headed coin (as we discussed on page 17), you have zero entropy.

Entropy uses logarithms. The logarithm of a number, in a base, tells you what power of the base gives that number. $2^3 = 8$. So the logarithm of 8 is 3, in base two. $3^2 = 9$. So the logarithm of 9 is 2, in base three. Also, notice you can add the powers:

$$(2 \times 2 \times 2) \times (2 \times 2) = (2 \times 2 \times 2 \times 2 \times 2) = 2^5$$

$$\Rightarrow 2^3 \times 2^2 = 2^{3+2} = 2^5$$

To find the entropy of your dataset, you'll be finding the logarithm of fractions.

> \\//
> ⸜⸝
> **Joe asks:**
> # How Do You Find Logarithms of Fractions?
>
> What power of two gives you 0.5? Think about some whole numbers first:
>
> $$2^3 = 2 \times 2 \times 2 = 8,\ 2^2 = 2 \times 2 = 4,\ 2^1 = 2,\ 2^0 = 1$$
>
> So, now let's try fractions. Let's find the power, p, for a fraction, say a half:
>
> $$2^p = \frac{1}{2}$$
>
> You know two times a half is 1, so you can do this:
>
> $$2 \times 2 \times 2 \times \frac{1}{2} = 2 \times 2 \times \left(2 \times \frac{1}{2}\right) = 2 \times 2 \times 1 = 2 \times 2 = 2^2$$
>
> $$\Rightarrow 2^3 \times 2^p = 2^2$$
>
> What do you add to 3 to get 2? -1. This gives the power, p, you need. Put it back in the equation and you get $2^3 \times 2^{-1} = 2^{(3-1)} = 2^2$
>
> This means the logarithm of 0.5 is -1, in base 2, since you just saw
>
> $$0.5 = \frac{1}{2} = 2^{-1}$$

Entropy uses the proportion, P, of your data with each category or feature value. The categories here are 0 or 1 for x and y. Technically, treating continuous numbers, which could be any real number, as discrete categories or whole numbers is unconventional, but it will work for this example. You multiply this by its logarithm in base two, and sum them all up (Σ, a capital letter sigma, means sum). Entropy is often represented by the letter H (possibly due to a confusion between the Greek letter η (eta) and an H[6]):

$$H = -\sum_{i=1}^{n} P(x_i) \log_2 P(x_i)$$

Since you're finding proportions or fractions of your data, your logarithms will be negative, so you flip the sign.

Entropy tends to be calculated in base two to give *bits*, although you can use a different base. If you use base two, the entropy of a binary class, like heads/tails or inside/outside of the bag, is between 0 and 1. For more classes, it might be greater than 1. Some later algorithms *normalize* this using a ratio, otherwise features with more categories tend to get chosen.[7] You do not need to normalize your turtle's data since you have two classes and a pretty equal number of x and y coordinates.

Let's calculate entropy with the four coordinates, (x=0, y=0, in), (x=1, y=0, in), (x=0, y=1, out), (x=1, y=1, out). When you use x, you have two possible values, 0 or 1, so need to do two calculations first. When x is 0, you have one of the two points in and the other out:

$$
\begin{aligned}
H(X = 0) \ &= \ -(P(\text{out}) \times \log P(\text{out}) + P(\text{in}) \times \log P(\text{in})) \\
&= \ -\left(\frac{1}{2} \times \log\frac{1}{2} + \frac{1}{2} \times \log\frac{1}{2}\right) \\
&= \ -(0.5 \times -1 + 0.5 \times -1) \\
&= \ -(-0.5 + -0.5) \\
&= \ -(-1) = \ +1
\end{aligned}
$$

When x is 1, you get the same value since half are in and half are out. To find the entropy of the split, sum the proportions of these two values in the whole set:

$$
\begin{aligned}
H(\text{split}) \ &= \ P(x = 0) \times H(x = 0) + P(x = 1) \times H(x = 1) \\
&= \ \frac{2}{4} \times 1 + \frac{2}{4} \times 1 = \frac{1}{2} + \frac{1}{2} = 1
\end{aligned}
$$

Lots of entropy. Now consider the y value instead. When y is 1, both points are out so your entropy calculation is:

6.   math.stackexchange.com/questions/84719/why-is-h-used-for-entropy
7.   https://en.wikipedia.org/wiki/Information_gain_in_decision_trees

$$H(Y=1) = -(P(\text{out}) \times \log P(\text{out}) + P(\text{in}) \times \log P(\text{in}))$$
$$= -\left(\frac{2}{2} \times \log\frac{2}{2} + \frac{0}{2} \times \log\frac{0}{2}\right)$$
$$= -(1 \times 0 + 0 \times \log 0)$$

By convention, log 0 is not defined, but you're trying to find $0 \times \log 0$ so you use 0 for this part of your sum. This gives you

$$= -(0+0) = 0$$

When y is 0, you also get 0 because the proportions are the same, though both points are now inside the bag. To find the entropy of this split, sum the proportions of these two values in the whole set:

$$H(\text{split}) = P(y=0) \times H(y=0) + P(y=1) \times H(y=1)$$
$$= \frac{2}{4} \times 0 + \frac{2}{4} \times 0 = 0 + 0 = 0$$

As you can see, you get much lower entropy if you use y.

To decide the best feature, you compare this with a baseline entropy you have across all your data without a split. For your set of four coordinates, you have two points in and two points out. Your baseline is, therefore:

$$H(\text{data}) = -(P(\text{in}) \times H(\text{in}) + P(\text{out}) \times H(\text{out}))$$
$$= -\left(\frac{2}{4} \times \log\frac{2}{4} + \frac{2}{4} \times \log\frac{2}{4}\right)$$
$$= -(0.5 \times -1 + 0.5 \times -1) = -(-1) = 1$$

You can then calculate *information gain* and pick the variable with the highest gain. This gain is the difference between the baseline entropy and entropy if you split on one attribute.

For x, you have 1 - 1 = 0, so no gain at all. For y, you have 1 - 0 = 1, so maximal gain. You've found the best split point. You already knew it was y, but this method applies to any dataset. Ready to find your paper bag?

Let's Find That Paper Bag

Using the saved data from the previous chapter, you first need to load it:

```
import pickle

with open("data", "rb") as f:
  L = pickle.load(f)
```

Your data is a list of lists. Each inner list has three items: ['x', 'y', 'out']. Your tree will predict the last item: 'out'. You'll provide a label for each column to help make readable rules from the tree which will be built up of sub-trees using split points. But first, you need to find the split points.

Find Split Points

You use information gain to find split points. This needs the proportions of data in each group. You can use the collections library to find counts of each value, which gets you most of the way there.

Try it on a list of numbers:

```
import collections
count = collections.Counter([1, 2, 1, 3, 1, 4, 2])
```

This gives you a Counter with the frequency of each item:

```
Counter({1: 3, 2: 2, 3: 1, 4: 1})
```

The keys are your numbers, and the values are the frequencies of each. The ratio you need is the frequency divided by the length of the list.

Information gain is the difference between baseline entropy and the entropy of each split. You, therefore, need an entropy function:

Decide/decision_tree.py
```
def entropy(data):
  frequency = collections.Counter([item[-1] for item in data])
  def item_entropy(category):
    ratio = float(category) / len(data)
    return -1 * ratio * math.log(ratio, 2)
  return sum(item_entropy(c) for c in frequency.values())
```

You use a Counter to find the frequency of each category, which is in the last column of your data, at index -1. You can then find the proportion or ratio (r), of each category by dividing by the length of your data. For each category, take the negative of the logarithm, in base 2, of this ratio multiplied by the ratio itself as you saw on page 21. Sum these to get the entropy of the data.

You can now find the feature with the most information_gain. Pull out the sample for each possible value of each feature and find the entropy. Best feature wins. Picking the best makes this a *greedy* algorithm which can lead to problems —if you choose what looks great now, you may miss something better later on. You will consider this later when you assess if this works. For now, be greedy:

Decide/decision_tree.py
```
def best_feature_for_split(data):
  baseline = entropy(data)
  def feature_entropy(f):
```

```
  def e(v):
    partitioned_data = [d for d in data if d[f] == v]
    proportion = (float(len(partitioned_data)) / float(len(data)))
    return proportion * entropy(partitioned_data)
  return sum(e(v) for v in set([d[f] for d in data]))
features = len(data[0]) - 1
information_gain = [baseline - feature_entropy(f) for f in range(features)]
best_feature, best_gain = max(enumerate(information_gain),
                            key=operator.itemgetter(1))
  return best_feature
```

You will use this to build your decision tree.

Build Your Tree

Using the collection Counter, you can call most_common(1) to determine the most frequent category used in the dataset. Then, you can use this to decide whether to make a leaf node for your decision tree:

Decide/decision_tree.py
```
def potential_leaf_node(data):
  count = collections.Counter([i[-1] for i in data])
  return count.most_common(1)[0] #the top item
```

This gives a tuple of the most common category and the count of items in this category. If all of your data is in one category, you can make a leaf node. If most of your data is in one category, you can also make a leaf node. To do this, you need to decide what counts as "most." To keep things simple, stick with 100% purity for now.

If you decide not to make a leaf node, you need to build a sub-tree instead. Make an empty dictionary {} and choose the best feature on which to split your data:

Decide/decision_tree.py
```
def create_tree(data, label):
  category, count = potential_leaf_node(data)
  if count == len(data):
    return category
  node = {}
  feature = best_feature_for_split(data)
  feature_label = label[feature]
  node[feature_label]={}
  classes = set([d[feature] for d in data])
  for c in classes:
    partitioned_data = [d for d in data if d[feature]==c]
    node[feature_label][c] = create_tree(partitioned_data, label)
  return node
```

If all of your data is in one category, return that category to make a leaf node. Otherwise, your data is *heterogeneous*, so you need to partition it into smaller groups by calling create_tree recursively.

You can now build a tree with some training data and labels. Next, you'll see how to use your tree to classify new data.

Classify Data

Although it's possible to print your tree to see the dictionary and then manually apply it to data, let's get your computer to do the work. The tree has a root node—the first key in the dictionary. If the corresponding value is a category, you've found a leaf node, and your job is done. If it's a dictionary, you need to recurse:

Decide/decision_tree.py
```
def classify(tree, label, data):
  root = list(tree.keys())[0]
  node = tree[root]
  index = label.index(root)
  for k in node.keys():
    if data[index] == k:
      if isinstance(node[k], dict):
        return classify(node[k], label, data)
      else:
        return node[k]
```

Remember the dictionary for the letter and number decision tree on page 17:

```
{'letter': {'a': 'good', 'b': 'bad'}}
```

For a new data point, ['b', 101], you get 'bad'. Why? The key of the root node, tree.keys()[0], is letter. You find the index of this label, getting 0. You data has 'b' at index 0, so you follow the 'b' branch of the sub-tree. You hit the value 'bad', so have your decision.

You can create and use decision trees. How do you make your tree into a ruleset?

Transform a Tree into Rules

You can use a graph library to build a visual representation of a tree, but for simplicity, you'll print the equivalent rules. You can adapt your classify function, noting the labels and corresponding values as you walk through the tree.

You need to start with an empty string and build up a rule saying if that label has a specific value, then you either check more if conditions or report the leaf node's value with a then. Like this:

Decide/decision_tree.py

```
def as_rule_str(tree, label, ident=0):
  space_ident = '  '*ident
  s = space_ident
  root = list(tree.keys())[0]
  node = tree[root]
  index = label.index(root)
  for k in node.keys():
    s += 'if ' + label[index] + ' = ' + str(k)
    if isinstance(node[k], dict):
      s += ':\n' + space_ident  + as_rule_str(node[k], label, ident + 1)
    else:
      s += ' then '  + str(node[k]) + ('.\n' if ident == 0 else ', ')
  if s[-2:] == ', ':
    s = s[:-2]
  s += '\n'
  return s
```

Let's see how good your trees are.

Did It Work?

It's time to check how well the classifier did. You can measure the performance of classifiers in various ways. You can find the *accuracy* by calculating the percentage correctly classified. For numeric data, you can use an *error function*, such as the *mean squared error* (MSE) which finds the average of the squares of the errors or difference between the predicted and actual values. For each problem, you need to decide how to test your algorithm.

How well has your decision tree performed? Try it on the four coordinates:

```
data = [[0, 0, False], [1, 0, False], [0, 1, True], [1, 1, True]]
label = ['x', 'y', 'out']

tree = create_tree(data, label)
print(as_rule_str(tree, label))
```

You get this rule:

```
if y = 0 then False.
if y = 1 then True.
```

The rule has picked the y coordinate to make a decision. This looks promising.

You can classify some points:

```
print(classify(tree, label, [1, 1]))
print(classify(tree, label, [1, 2]))
```

Your tree says the point (1, 1) is outside of the paper bag. However, it does not know what to do with (1, 2), so you receive None back. You knew a tree built from the four training points would not cope with other points. Try your algorithm on the five training points you considered earlier on page 20:

```
data = [[0, 0, False],
        [-1, 0, True],
        [1, 0, True],
        [0, -1, True],
        [0, 1, True]]
label = ['x', 'y', 'out']
tree = create_tree(data, label)
category = classify(tree, label, [1, 1])
```

Does it decide the unseen coordinate is outside the bag? Yes, it does. Success. The rule looks like this:

```
if x = 0:
  if y = 0 then False, if y = 1 then True, if y = -1 then True
if x = 1 then True.
if x = -1 then True.
```

By using more data with a greater variety of possible features, you're less likely to get a point your tree cannot classify. You'll still have the same problem if you ask about a point with a coordinate value you didn't see in the training data. Supervised learning algorithms cannot guess what to do with entirely new data that is unlike their training data. You can make them *extrapolate* instead if that's appropriate for your problem. Then they can step beyond the training data.

For category data, you can't extrapolate. For numeric data, you can make your tree partition data smaller or larger than a feature value, say the median. This allows your classifier to extrapolate beyond the minimum or maximum values it trained on. For this problem, you know you're after a square paper bag, so any points left, right, above, or below the edge points are outside. You can use this to make neat rules.

Generate a decision tree for the data you saved in the previous chapter and then print the rules. Note, however, the rules can get long since they state what to do for every x or y coordinate. Here's a small sample of the rules generated from the data points on the squares in the previous chapter:

```
if x = 3.0 then False.
if x = 20.0 then False.
if x = -17.0 then False.
if x = -45.0 then True.
if x = -45.0 then True.
if x = -46.0 then True.
```

The points were at the corners of the squares, so whether a point is inside or outside the bag, your x and y coordinates will have the same value, though can have opposite signs. The decision tree has picked the x coordinate for each rule since it saw this first.

You can also see -45.0 twice when you print the rule because the numbers are rounded to a single decimal place. You can end up with one rule per data point if you aim for 100% purity. By merging back—or *pruning*—these nodes to the parent sub-tree, you drop some purity, but this can avoid *overfitting* the data. Overfitting tends to give high accuracy on the training data, but your algorithm does poorly on new data. For this example, you don't get this problem, but you do get lots of rules. Let's prune these back to get neater rules.

How to Prune Your Rules

You know the bag was square so you can use the smallest and largest x and y coordinates inside the bag to describe its edges. This gives you a pruned ruleset indirectly.

If you find matching x and y values in your training data and scan along these pairs, you sweep up the diagonal as you can see:

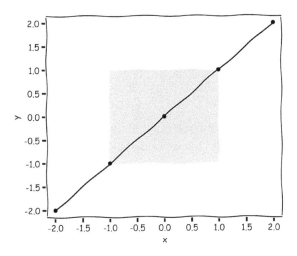

The set intersection function finds matching values, so use this to find these points on the diagonal. If you classify these points, you can find the smallest and largest inside the paper bag. These tell you (min_x, min_y) and (max_x, max_y). Like this:

Decide/decision_tree.py
```python
def find_edges(tree, label, X, Y):
  X.sort()
  Y.sort()
  diagonals = [i for i in set(X).intersection(set(Y))]
  diagonals.sort()
  L = [classify(tree, label, [d, d]) for d in diagonals]
  low = L.index(False)
  min_x = X[low]
  min_y = Y[low]

  high = L[::-1].index(False)
  max_x = X[len(X)-1 - high]
  max_y = Y[len(Y)-1 - high]

  return (min_x, min_y), (max_x, max_y)
```

Some shapes work better than others. If you only had a horizontal or vertical line, you cannot work out the full width or height of the bag. For my random spirangles, I got 100 or so data points, with over 90 rules. If you find the edges, this condenses down to four points, to two decimal places, of:

```
(-33.54, -33.54), (34.32, 34.32)
```

The edges were at (-35, -35), (35, 35). It's close, but not perfect. Your numbers might differ since this was randomly generated. If you use data from your squares instead, you have about 200 data points, more evenly spread. This finds (-35, 35, -35, 35), give or take some rounding. This is even better than the spirangles. You can use these points to form a much neater rule, which covers any values:

```
if x < 35.0 or x > 35.0 or y < 35.0 or y > 35.0 then True
else False
```

That's better. You won't get None back now. A combination of the right training data and some domain knowledge helped you locate the paper bag. In this case, you can tell the decision tree or rules are correct because I reminded you where the paper bag was. For other problems, you can't be sure your rules are correct.

For real-world data, you should try *validating* your trees against unseen data, tweaking any parameters; for example, the purity required at a leaf node, or by pruning rules back. Once you're happy with your model, *test* it on some more unseen data. The "Train, Validate, and Test" pipeline is common in machine learning.[8] You should always test your code.

---

8. en.wikipedia.org/wiki/Training,_test,_and_validation_sets

Over to You

In the previous chapter, you escaped a paper bag; in this chapter, you used supervised machine learning to locate the edges of the paper bag. To keep things simple, you used category data, treating each coordinate as a specific value, rather than a number from a possible range. Because you knew what shape the paper bag was, you were able to reformulate the decision tree using numeric ranges. Other decision tree algorithms, such as C4.5 or random forests can use numeric data directly.[9] Their details differ, but they still start with training data, dividing it up to build a model.

You can tweak your edge-finding algorithm to cope with rectangles instead of squares. Decision trees tend to carve up the *input space* into rectangular regions, like

```
(-35 < x < 35) and  (-35 < y < 35)
```

Other approaches can make *oblique trees*, making linear combinations like

```
(-35 < 2x - y < 35)
```

These still have straight lines on the *decision boundaries*, though they slope. Other algorithms, such as *support vector machines*, can find curved or even more complicated dividing lines.

In the next chapter, you'll learn how to use a *genetic algorithm* by firing virtual cannonballs at various speeds and angles from inside a paper bag. Some will get out, and your algorithm will learn how to get more out over time. Genetic algorithms have a long history and have been used to solve all kinds of different problems. They start with randomly generated solutions and iteratively improve. Many machine learning algorithms take a similar approach so that you will get a feel for the essence of many algorithms. Let's fire some cannons.

---

9.   C4.5/C5.0 is Quinlan's follow to ID3. See www.rulequest.com/see5-info.html

CHAPTER 3

Boom! Create a Genetic Algorithm

In the previous chapter, you used saved data points a turtle visited as he escaped a paper bag to build a decision tree. By splitting the data on attributes, specifically x and y coordinates, the tree was able to decide whether new points were inside or outside the original paper bag. This is one way to predict values for new data. There are many other ways to make predictions. So many, in fact, that they could fill several volumes with the different approaches. One is enough to get a feel for this flavor of AI. Let's try something completely different. Rather than predicting the future or outcomes, can you find combinations or suitable inputs to solve problems?

- How do you split your investments, maximizing your pension and avoiding buying shares in companies you have moral qualms over?

- How do you create a timetable for classes, making sure there are no clashes, and all the classes get taught and have an assigned room?

- How do you make a seating plan for a wedding, making sure each guest knows someone else at the table and keeping certain relatives as far apart as possible?

These seem like very different problems. The investments will be amounts of money in different schemes. The timetable will tell you what happens when and where. The seating plan will be seat numbers for each guest. Despite these differences, they have something in common. You have a fixed number of investments that need values or classes or guests to arrange in a suitable order. You also have some constraints or requirements, telling you how good a potential solution is. Any algorithm returning a fixed-length array organized or populated to fulfill conditions will solve the problem.

For some problems, you can work through the options or use mathematics to find a perfect solution. For one fixed-rate bond in a portfolio, the mathematics to find its future value is straightforward. With one class, one teacher and one room, there is only one possible timetable. For a single guest at a wedding, the seating plan is apparent. For two or three guests you have more options, but can try out each and decide what to do. Once you have 25 guests, there are 15,511,210,043,330,985,984,000,000 possible arrangements.[1] Trying each of these, known as *brute force*, against your constraints will take far too long. You could reject a few combinations up front but will still be left with far too many to try. All you need is a list of seat numbers for 25 people. You could try a few at random and might get lucky, but might not. Ideally, you want a way to try enough of the possible arrangements as quickly as possible to increase the chance of finding a good enough seating plan (or timetable, or investment).

There is a machine learning or evolutionary computing method called a *genetic algorithm* (GA) that is ideal for problems like this. A GA finds a solution of fixed length, such as an array of 25 guests' seat numbers, using your criteria to decide which are better. The algorithm starts with randomly generated solutions, forming the so-called initial population, and gradually hones in on better solutions over time. It is mimicking Darwinian evolution, utilizing a population of solutions, and using the suitability criteria to mirror natural selection. It also makes small changes, from time to time, imitating genetic mutation.

The algorithm makes new populations over time. It uses your criteria to pick some better solutions and uses these to generate or *breed* new solutions. The new solutions are made by splicing together *parent* solutions. For investments of bonds, property, foreign exchange and shares, combine bonds and property from one setup with foreign exchange and shares from another, and you have a new solution to try out. For seating plans, swap half of one table with half of another, or swap parts of two seating plans. You might end up with the same seat used twice, so you need to do some fixing up. There are lots of ways to splice together arrays. The GA also mutates elements in the solution from time to time, such as swapping two people's seats. This can make things worse—splitting up a couple might not be good—but can make things improve too. Nonetheless, this keeps variety in the solutions thereby exploring several of the possible combinations.

There are many ways to select parent solutions, tournament and roulette wheel being common. We'll use roulette wheels in this chapter and try tournaments later in Chapter 9, *Dream! Explore CA with GA*, on page 163. Once we've created a GA, we'll have a look at mutation testing to evaluate unit tests, emphasizing mutation as a useful general technique. This chapter adds

---

1. www.perfecttableplan.com/html/genetic_algorithm.html

to your fundamental concepts, encourages you to question your options, and helps you build a simple genetic algorithm.

Imagine a paper bag with a small cannon inside, which can fire cannonballs at different angles and velocities. If your mission is to find a way to fire these cannonballs out of the bag, how would you go about doing this? You have a few options:

- Work through the mathematics to find suitable ranges of angles and velocities. This is possible for this problem, but won't show us how GAs work.

- Use brute force to try every combination, but this will take ages.

- Build a genetic algorithm to find pairs of angles and velocities that send the cannonballs out of the bag.

Your Mission: Fire Cannonballs

Let's create a genetic algorithm for firing virtual cannonballs out of a paper bag. Let's see how cannonballs move when fired, and start thinking about which paths are better. This will tell us the criteria for the GA. There are two ways these cannonballs can move: straight up or at an angle.

When fired at an angle, cannonballs travel up, either left or right, eventually slowing down due to gravity, following a parabola. When fired straight up at 90 degrees, a similar thing happens. However, instead of a parabola, they come straight down. Cannonballs fired fast enough go into orbit, or reach escape velocity, which is certainly out of the paper bag, but hard to draw.

The trajectories of a few cannonballs are shown in the next figure. They start from an invisible cannon located at the bottom/middle of a gray bag. One cannonball travels up a little, then falls to the bottom of the bag and rolls along. Another two go up and stick to the edge of the bag. Finally, one manages to escape the bag by going high enough and fast enough:

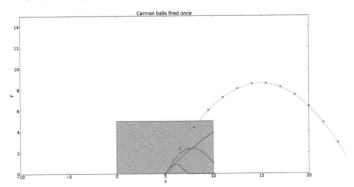

The higher up a cannonball gets at the edge of the bag, the better the angle, velocity pair. Any cannonball over the bag height at the edge escapes. You can use this height as the criteria, or fitness function. Let's see what equations to use to map these trajectories.

The coordinates of a cannonball, or any ballistic, at a specific time (t) can be found using an initial velocity (v) and an angle (θ). With the velocity in meters per second, and the angle in radians, on a planet that has gravity (g)—which, here on Earth is ~9.81 meters per second squared—the coordinates (x, y) of a cannonball (t) seconds after firing, are found using these equations:

$$x = vt\cos(\theta)$$

$$y = vt\sin(\theta) - \frac{1}{2}gt^2$$

\\/ Joe asks:

ʒʃ What's a Radian?

The trigonometry functions in Python use radians. They start with 0 on the horizontal axis and move around counter-clockwise; see the figure. There are 2π radians in a full circle, which equals 360 degrees. Therefore, 1 radian is equal to $180/\pi$ degrees, and 1 degree is equal to $\pi/180$. Radians were introduced to make some mathematics easier. You can use the radians conversion function or use radians directly.

The figure shows angles in degrees and radians, starting with 0 on the right and moving counter-clockwise:

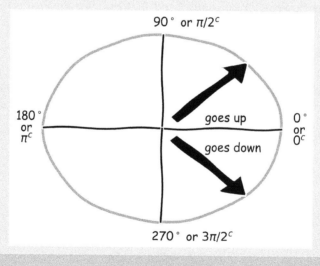

The first step in using a GA is encoding your problem. The cannonballs need pairs of numbers—angles and velocities—for their trajectories. Dividing up investments into bonds, property, foreign exchange, and shares finds four numbers—how much in each. Finding a seating plan can use an array with one element per guest. Other problems find letters or strings of bits. Any fixed-length solution is amenable to GAs.

To get started, the GA creates some random pairs of numbers to plug into these equations. The wedding guest problem would also generate some random seating plans to get started. Two guests would also be a pair, while more guests needs a list or array. Armed with pairs of angles and velocities, the GA creates new pairs. It selects some better pairs as *parents* to create or *breed* new attempts. Using the *fittest* pairs to create new solutions to a problem alludes to Darwin's theory of evolution.[2] That's why a GA is an evolutionary algorithm. Two parents are spliced together to make a new solution. Some numbers come from one parent and some from the other. This is called *crossover*, using a simplified model of genetic recombination during sexual reproduction where parents' strands of DNA splice together when living creatures breed.

The GA also uses *mutation*, in this example making either the angle or velocity a bit bigger or smaller. The crossover and mutation operations keep some variety in the population of angle, velocity pairs. Remember, you don't have time to try all the numbers, but want to try as many as possible to increase the chance of finding something that works. The crossover and mutation operations take care of that for us.

How to Breed Solutions

You know the solutions are pairs of numbers, which you need to initialize. You can weed out a few numbers with a little thought since they are bound to fail. Armed with a handful of initial attempts, you need a way to pick better ones to breed even better solutions. Once you have a clear idea of the backbone of the algorithm, you can implement the crossover operator to breed new solutions and mutate these from time to time in Python (either 2.x or 3.x), as described in *Let's Fire Some Cannons*, on page 40. Feel free to write the code as you read, or start thinking about how you want to implement this. Be sure to write some unit tests so that you can use these for *Mutation Testing*, on page 52.

---

2. https://en.wikipedia.org/wiki/Survival_of_the_fittest

Starting Somewhere

The GA needs pairs of velocities and angles to get started. It can pick any numbers, but some are going to fail. The GA will weed these out, but you can help it along by sharing some things you already know about the problem. For a seating plan, you could ensure couples sit together. For the cannonballs, when the velocity is zero, the cannonball won't go anywhere, so the velocity needs to be a number greater than zero.

What about the angle? Anything more than a full circle is like spinning the cannon around a full circle, plus a bit more—the outcome is the same as just tilting the cannon by the bit more. In fact, anything less than 0 or greater than half a circle fires downwards, so your angles should be in the first half circle: between 0 and π.

How many pairs should you generate? At least one, maybe a few more. The idea is to start small and add more if you think it's needed. Once you have a list of pairs, you come to the clever part of the GA. It runs for a while, picking some of the better pairs from last time to combine into new pairs.

For a While...

The first set of solutions of velocity/angle pairs are created randomly, giving a variety of solutions, some better than others. This is the first *generation*. The GA will use these to make better solutions over time. A GA is a type of guided random search, or *heuristic search*, using these to find improved solutions, using a loop:

```
generation = random_tries()
for a while:
  generation = get_better(generation)
```

This form crops up again and again. This is often described as a random *heuristic* search. Something random happens and is improved guided by a heuristic, *fitness function*, or some other way to refine the guesses.

There are many ways to decide when to stop searching. You can stop after a pre-chosen number of attempts or *epochs*; you can keep going until every pair is good enough, or you can stop if you find one solution that works. It all depends on the problem you are trying to solve. For a seating plan, you could let it run for a few hours and have a look at the suggestions. There is no *one true way* with machine learning algorithms. Many variations have official names, and you might be able to think up some new variations. Do a search for "genetic algorithm stopping criteria" on the internet for further details. What you find will include talk of NP-hard problems, probabilistic upper bounds, and convergence criteria. Before learning about those, let's create a GA since right now, you have a paper bag to code your way out of!

How to Get Better

The GA uses a heuristic to assess items in the current population or generation. Some of the better pairs are used to make a new generation of solutions. The driving idea here is the survival of the fittest, inspired by Darwinian evolution. Finding fitter parents might make fitter childeren.

To select fitter parents, the GA needs a way to compare solutions. For a seating plan, you can check how many criteria are satisfied. For an investment, you can check how much pension you will get. Which are the fittest (angle, velocity) pairs? There are options. Let's consider two approaches: either find pairs that get out of the bag, or decide a way to rank the pairs by finding a score for how well they did.

For the first approach, you can follow the arc of each cannonball and see which ones leave the bag. Does a ball hit an edge, land back inside the bag, or get out? For the second approach, you can do some math and figure out how far up the ball went, and how close to the edge it was. Look back at the cannonball paths on page 35 to get an idea of the different trajectories.

Both approaches provide a way to measure the viability of a solution. This is the heuristic part of the algorithm, and it is what guides the algorithm toward better solutions. With the first approach, you return a boolean based on whether or not the ball made it out of the bag. With the second approach, you return the y value when the ball is at the edge of the bag. Then you assign a score based on that value: the closer it was to escaping, the better the score.

So which is better? Let's unpack things a bit. If a cannonball nearly gets out but doesn't quite make it, the first option will brutally declare it a failure. With the second option, however, this information will be passed on to future generations. The mutation function might even nudge these toward success.

In the next section, we'll use the height for fitness. Feel free to compare this solution to the more brutal method of using a boolean. Any GA you create needs a heuristic or fitness function, so always needs a bit of thought. A fair bit is hiding behind the random start and gets better in a loop! There are still a few decisions to make, so let's consider these. After that, *Let's Fire Some Cannons*, on page 40 walks through an implementation to select parents, breed solutions and mutate them from time to time, giving you a working GA.

Final Decisions

You now have an overview of a genetic algorithm—a random setup and a loop, with the all-important fitness function—but the GA needs some magic numbers to run. How many epochs? One epoch is fine for a trial run, but you need more to see improvement. Instead of a pre-chosen number of epochs, you can wait for an average fitness, minimum fitness, or until one or all pairs work. Looping for a pre-specified number of times is straightforward, so try that first. Of course, you can tweak this and try other options too.

How big should a population be? This number can make a big difference. Imagine if you just had one. How will it breed? If you have a couple, and you used a pass/fail *fitness function*, what happens if neither is good? You can try something random again, but then you're back to the start. Going to the other extreme is a waste too. Trying 1,000,000 solutions when only 10 work is a waste. It is best to start with a small number first, to flush out any problems, and to increase parameters only if needed.

Let's try twelve solutions and ten epochs, giving the backbone of the algorithm:

```
items = 12
epochs = 10
generation = random_tries(items)
for i in range (1, epochs):
  generation = crossover(generation)
  mutate(generation)
display(generation)
```

All genetic algorithms have the same core shape; try something random, then loop around trying to improve. In fact, many machine learning algorithms look like this. When you discover a new machine learning algorithm, look for how it starts and where it loops. The differences come from how solutions improve. Let's fill in the implementation details for crossover, mutation, and draw the arcs of the cannonballs with Matplotlib.

Let's Fire Some Cannons

The GA starts with a generation of random_tries. The GA selects parents from each generation to breed new solutions by crossover. It then has a new generation and will perform mutation on a few of the pairs to keep some variety. Crossover picks angles and velocities from the parents, so mixes things up a bit, but mutation ensures more variety, by changing the numbers. Crossover could splice together two seating plans but would need to deal with two guests ending up in the same seat. Mutation could swap two guests. These two operations depend on the problem at hand.

You can try crossover or mutation by themselves, but the combination allows the GA to explore more and makes it more likely to find a good solution more quickly. Let's begin by making the first generation, then create and use a fitness function to crossover the angle-velocity pairs, making new solutions. You can then mutate these once in a while, and your GA will find some suitable candidates.

Random Tries

To get started, the GA needs a population of candidate solutions. A list fits the bill, so make a list and append pairs of random numbers; theta for the angle and v for velocity. Make the velocity greater than zero, or the cannonball will not move. For the angle, anything greater than 0^c and less than π^c is sensible, otherwise you are firing into the ground.

Import the random package and fill the list:

Boom/ga.py
```
def selection(generation, width):
    results = [hit_coordinate(theta, v, width)[1] for (theta, v) in generation]
    return cumulative_probabilities(results)
```

You now have a list of possible solutions. For a seating plan, a few permutations of seat numbers is an excellent place to start. For investments, a few randomly selected amounts in each asset, totaling the amount you have to invest, would work. Each problem needs its own setup, but in essence, creates a list that's the right size for the problem. Some solutions will be better than others. The genetic algorithm will improve these over time by selecting some good parents and letting them breed.

Selection Process

How do you select parent solutions? A genetic algorithm tends to pick better, stronger, fitter solutions for breeding, as happens in Darwinian evolution. This means the fitter solutions can pass on their genes to future generations. In this case, the genes are angles and velocities.

Armed with a fitness function, the GA can decide which solutions are better. Breeding the best each time would end up with lots of identical attempts, which might not be good enough to escape the paper bag. *Using the Fitness Function*, on page 43 demonstrates how to pick some of the better solutions, without always going for the very best. Whichever selection algorithm you use, you need a way to choose between solutions. This comes from a fitness function.

Creating the Fitness Function

The arcs of the cannonballs let you pick a few good solutions by eye. But how can the algorithm make that decision? You considered two approaches earlier. It can check whether or not cannonballs escaped, but that might miss some solutions that get close. Finding out how high up a cannonball gets when it is at the side of the paper bag allows solutions that are close but not perfect. The GA can improve on these. Let's figure out this height.

Recall how to find the x-coordinate of a cannonball:

$$x = vt\cos(\theta)$$

Since you know how wide the paper bag is, you can work out when it gets to the edge, and then use the equation for y to see how high up it gets at time t. Suppose the cannon is positioned in the middle of a bag that is 10 units wide, as it was in the arcs considered on page 35. If the bottom left of the bag is (0, 0), then a cannonball at 0 or +10 across is at the edge. Since it started in the middle, it has traveled 0.5 * 10 = 5 units left or right when it gets to the edge. If it goes right, anything under 90 degrees, it has traveled 5 units across when it gets to the edge, so the time it took is

$$t = 5/(v \times \cos(\theta))$$

You know v and theta, so can calculate t. If it goes left, anything over 90 degrees, it travels -5 units horizontally, so you can do a similar sum using -5 instead of 5.

Having found t, find the height it gets to at the edge of the bag using the equation for y:

$$y = vt\sin(\theta) - \frac{1}{2}gt^2$$

To find the coordinates of cannonballs at the bag edge, let's pull this together in code:

Boom/ga.py
```python
def hit_coordinate(theta, v, width):
  x = 0.5 * width
  x_hit = width
  if theta > math.pi/2:
    x = -x
    x_hit = 0
  t = x / (v * math.cos(theta))
  y = v * t * math.sin(theta) - 0.5 * 9.81 * t * t
  if y < 0 : y=0.0
  return x_hit, y
```

You can use this to make a fitness function returning a boolean by checking whether or not the cannonball goes over the bag height. Alternatively, you can use the y value it returns as a fitness. Higher values will be better. An escaped function will be useful, to pick out any cannonballs that make it out of the paper bag:

Boom/ga.py
```
def escaped(theta, v, width, height):
  x_hit, y_hit = hit_coordinate(theta, v, width)
  return (x_hit==0 or x_hit==width) and y_hit > height
```

Using the Fitness Function

Armed with a list of potential solutions, and a fitness function, the GA can now select parents. It could take the best two, but to give your GA a chance to explore a variety of possible solutions, you need to be slightly less elitist. You can pick the top few by sorting or *ranking* the population and pick two at random from the top few. This will tend to drive your solutions down one path; mutation will give some variety, so this can work. You can pick a few solutions, say three or four, and let them compete, using the winner of these *tournaments* as parents. Chapter 9, *Dream! Explore CA with GA*, on page 163 builds another GA for more than two numbers, using a tournament. Ranking and tournaments need a number—how many to pick or get to compete. There is another approach that doesn't need this extra parameter.

This method, called *proportionate selection*, tends to choose better solutions by treating the fitness like a probability—bigger numbers are more likely. The solutions do not need to be sorted so it can be quicker. A simple and efficient way to do this is *roulette wheel selection*. This has drawbacks—for example, not coping with negative fitness values, or converging early if one solution is much better than others but still not solving the problem. Nonetheless, it can succeed and is frequently used. Let's see how this selection method works.

A normal roulette wheel has equal sections, so each is equally likely. If the sections vary in size, the roulette ball is more likely to land in the larger areas. You can use the fitness value, here the y coordinate, to make different sized sections. The fitter values are more likely to be picked, but any section might get picked. This ensures some variety, compared to only ever choosing the best.

Let's use the y value from four arcs similar to those you saw earlier on page 35 to make a roulette wheel. We'll use 0 if it doesn't get to the edge of the bag. The fitness tells you what proportion of the roulette wheel a solution uses. You need the cumulative sum of the fitness to calculate the proportion as shown in the table on page 44.

Solution	Fitness	Sum(fitness)
1	15	15
2	1	16
3	8	24
4	6	30

Now you can draw these as a pie chart (or uneven roulette wheel), with the fitness giving the slice size. They sum to 30, so the first slice is 15/30, or half a circle. The next is only 1/30 of the circle. Then 8/30 and finally 6/30. If you sketch this as a roulette wheel it will look like this:

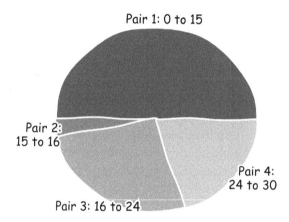

Since they sum to 30, you can pick a random number between 0 and 30 to select a pair—it will be like a roulette ball falling in one of the sections; under 15 is the first solution, between 15 and 16 is the next, and so on. The first solution is more likely to be picked since it has a bigger slice.

In code, find the hit coordinates, sum the heights, and then choose a solution, like this:

Boom/ga.py
```python
def cumulative_probabilities(results):
  #Could use from itertools import accumulate in python 3
  cp = []
  total = 0
  for res in results:
    total += res
    cp.append(total)
  return cp #not actually a probability!
```

```
def selection(generation, width):
  results = [hit_coordinate(theta, v, width)[1] for (theta, v) in generation]
  return cumulative_probabilities(results)

def choose(choices):
  p = random.uniform(0, choices[-1])
  for i in range(len(choices)):
    if choices[i] >= p:
      return i
```

The selection function appends the y-coordinate where each ball hits the bag edge to a list. The cumulative_probabilities function then stacks up the running total of these, giving you your roulette wheel. To be probabilities, the fitnesses ought to add up to 1. However, you don't need to scale them, if you pick a number between 1 and the total. You shouldn't call them probabilities if you do this, but the algorithm works out the same. The choose function spins the roulette wheel by picking a random number and seeing which solution it corresponds to.

Crossover

You now have a population and a way to pick better parents. Armed with two parents, breed new solutions in a crossover function. Not all genetic algorithms use two parents; some just use one, and there are others that use more. However, two is a conventional approach:

Boom/ga.py
```
def crossover(generation, width):
  choices = selection(generation, width)
  next_generation = []
  for i in range(0, len(generation)):
    mum = generation[choose(choices)]
    dad = generation[choose(choices)]
    next_generation.append(breed(mum, dad))
  return next_generation
```

Notice this makes a new generation of the same size as the last generation. You don't need to do this; you can grow or shrink this at each iteration. Let's keep it simple for now, with a fixed number of items. Also notice the last_generation won't be used again. In effect, it dies off. You can select a couple of good solutions and let them live on. This is called *elitist* selection. Alternatively, you can kill off the worst few and keep the rest, adding enough children to make a full population. Entirely replacing the whole population each time is straightforward, so let's implement that. Let's see what happens in the breed function.

Crossover splices together solutions. Imagine four potential solutions of velocity and angle pairs. The selection algorithm will pick two of these. A new child solution is bred from the parent's genes, like this:

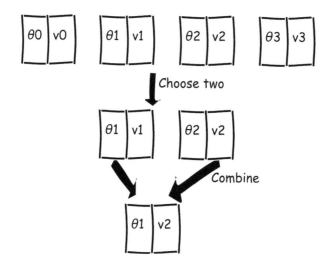

In order to breed, the information is split—half from one parent and half from another. Since there are two bits of information, each child will have a velocity from one parent and an angle from the other. There is only one way to split this. In general, when the solution contains more information, the genetic algorithm needs to be a bit smarter. It can interleave two solutions together or swap two chunks. For this problem, the simplest approach works:

Boom/ga.py
```
def breed(mum, dad):
    return (mum[0], dad[1])
```

In this example, the parents produce a single child. You can return both possible pairings and append those to your new generation instead. It's up to you!

Mutation

You now have a population of possible solutions and can breed the next generation using crossover. Your fitness function will tend to pick better solutions. It may throw up a few surprises, but it might converge to one solution referred to as *stagnation*. If this solves your problem, that's OK. However, there might be a better solution. To avoid this, a genetic algorithm adds mutation. In evolution, mutation helps with natural selection—things mutate, and successful mutations survive, eventually breeding new, stronger, fitter species.

How do you add mutation to your code? You have solutions with a velocity and an angle. Do you change one of these values or both? How can you change these a little? Or a lot? How often should you change them? If you want to experiment, you can control this from the outside by sending in a parameter. Some genetic algorithms keep a constant mutation rate—once per epoch, or less (or more)—but the same rate over all of the epochs. Some dampen it off over time. To find out more look up *genetic algorithm mutation rate*, but not yet! Let's code it first.

Let's potentially change everything in a population, to keep it general. Suppose you want something to happen on average, one time out of ten. One approach is to get a random number between 0 and 1 and do that "something" if you get less than 0.1. This is called *probabilistic* mutation. If you mutated values every time, this would be *deterministic*. Mutate either (or both) value(s) whenever you draw a random number less than 0.1.

How do you mutate the value? It's traditional to add or subtract a small amount or scale (multiply) a little for real numbers. Some GAs have solutions composed of bits, so mutation flips bits. A seating plan can swap guests. For real numbers, you can change by constant or random amounts, or try something more advanced.

Let's add a random number to the angle, but only use this if it stays between 0 and 180 degrees. Mutation can give rise to bad solutions which you can kill off immediately. For the velocity, scale by something random between 0.9 and 1.1, cunningly avoiding the problem of potentially getting zero; any mutated velocity is then OK.

Boom/ga.py
```python
def mutate(generation):
  #Could just pick one e.g.
  #i = random.randint(0, len(generation)-1)
  # or do all
  # or random shuffle and take top n
  for i in range(len(generation)-1):
    (theta, v) = generation[i]
    if random.random() < 0.1:
      new_theta = theta + random.uniform(-10, 10) * math.pi/180
      if 0 < new_theta < 2*math.pi:
        theta = new_theta
    if random.random() < 0.1:
      v *= random.uniform(0.9, 1.1)
    generation[i] = (theta, v)
```

Let's pull this together and see if it works.

Did It Work?

You can run the genetic algorithm like this:

```
Boom/ga.py
def fire():
  epochs = 10
  items = 12
  height = 5
  width = 10

  generation = random_tries(items)
  generation0 = list(generation) # save to contrast with last epoch

  for i in range(1, epochs):
    results = []
    generation = crossover(generation, width)
    mutate(generation)

  display_start_and_finish(generation0, generation, height, width)
```

What you do in the display function is up to you. *Plotting*, on page 48 plots the arcs of the cannonballs from the first and last generation, so shows any improvement on the initial random attempts. There are other ways to assess your solutions. *Counting*, on page 51 discusses a few things worth counting at each epoch to see if the GA is learning. For any algorithm it's worth writing some tests to make sure your code does what you intended. You'll see a way to assess these too, as described in *Mutation Testing*, on page 52. Elements of a genetic algorithm can be used in many situations. We'll round this section off with *Other GA Variations*, on page 53, touching on genetic programming and fuzzers.

Plotting

Let's plot the arcs of the cannonballs using Matplotlib. If you don't have it installed, use the Python package manager pip: pip install matplotlib.[3]

Import the library. People usually shorten it to plt, like this:

```
import matplotlib.pyplot as plt
```

To display your results, you needs some axes. For one plot, use plt.axes(). To draw two plots, use fig.add_subplot(2, 1, 1) for the first of two rows of subplots in one column, and fig.add_subplot(2, 1, 2) for the second.

---

3.   matplotlib.org/faq/installing_faq.html#how-to-install

Decide your bag's height and width and send these in along with the generation you want to display:

Boom/ga.py
```
Line 1   def display(generation, ax, height, width):
             rect = plt.Rectangle((0, 0), width, height, facecolor='gray')
             ax.add_patch(rect)
             ax.set_xlabel('x')
      5      ax.set_ylabel('y')
             ax.set_xlim(-width, 2 * width)
             ax.set_ylim(0, 4.0 * height)
             free = 0
             result = launch(generation, height, width)
     10      for res, (theta, v) in zip(result, generation):
               x = [j[0] for j in res]
               y = [j[1] for j in res]
               if escaped(theta, v, width, height):
                 ax.plot(x, y, 'ro-', linewidth=2.0)
     15          free += 1
               else:
                 ax.plot(x, y, 'bx-', linewidth=2.0)
             print ("Escaped", free)
```

First, draw the bag using Matplotlib's Rectangle using the height and width. Put the bottom left at (0, 0) and leave some width to the left and right of the bag when you set the plot width, as shown on line 6. Leave some space at the top for the cannons balls to fire over the bag—say four times the height, as shown on line 7.

The escaped function tells you if a cannonball escaped the bag, so you can show the difference between good and bad solutions. The code shown uses red circles 'r0' on line 14 for good solutions and blue crosses 'bx' on line 17 for other arcs.

The launch function uses the velocity and angle of each solution in the population to find these arcs. It starts cannonballs at (0.5 * width, 0), making a list of points every second. It stops if the balls hit the bag's edges or goes on for a few more points if they escape:

Boom/ga.py
```
def launch(generation, height, width):
  results = []
  for (theta, v) in generation:
    x_hit, y_hit = hit_coordinate(theta, v, width)
    good = escaped(theta, v, width, height)
    result = []
    result.append((width/2.0, 0.0))
    for i in range(1, 20):
      t = i * 0.2
      x = width/2.0 + v * t * math.cos(theta)
      y = v * t * math.sin(theta) - 0.5 * 9.81 * t * t
```

```
    if y < 0: y = 0
    if not good and not(0 < x < width):
      result.append((x_hit, y_hit))
      break
    result.append((x, y))
  results.append(result)
return results
```

You can count how many are good, or plot the best at each epoch. Plotting the initial attempts and the final attempts using the sub-plot feature of Matplotlib works well. This shows if the GA made better solutions in the end or not:

Boom/ga.py
```
def display_start_and_finish(generation0, generation, height, width):
  matplotlib.rcParams.update({'font.size': 18})
  fig = plt.figure()
  ax0 = fig.add_subplot(2,1,1) #2 plots in one column; first plot
  ax0.set_title('Initial attempt')
  display(generation0, ax0, height, width)
  ax = fig.add_subplot(2,1,2) #second plot
  ax.set_title('Final attempt')
  display(generation, ax, height, width)
  plt.show()
```

The results may vary, but they do *almost always* fire out of the bag. If a ball goes to the right initially and escapes, the rest tend to follow in that direction. If they go to the left, then so do the others. Sometimes the cannonballs go in both directions. The next set of images show a few outcomes. Sometimes most of the parabolas end up going right:

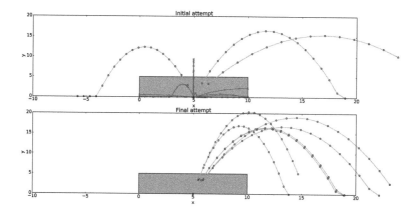

Sometimes, most go left:

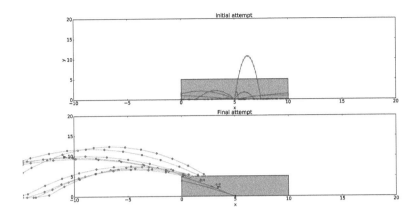

Once in a while, you get a mixture of right and left:

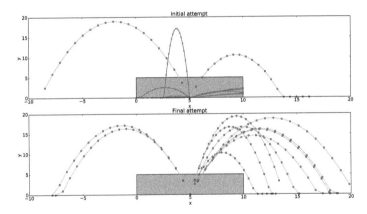

Counting

Whether or not you draw plots is up to you. You can make an animation, which updates at each epoch, and see if the genetic algorithm *learns* better solutions. Alternatively, pull out the best and worse solution, each time, and plot these. You can wait until you are done, and just plot the succeeding arcs from discovered (velocity, angle) pairs. Or, you can avoid plotting completely. Each epoch can show a count of how many escaped. There are several ways to assess how good the GA is.

Do your cannonballs all tend to end up on the right or the left? Did you ever get some going over either side? Just counting the successful shots will miss this. It will be clear from the plots.

You probably thought of many other questions and things to try. Great! You are now ready to explore more. Before wrapping up this chapter, let's see a use for mutations by applying mutation testing to unit tests for your GA and touch on a more general idea than genetic algorithms: *genetic programming*.

Mutation Testing

One way to assess code is by writing tests. Mutation testing provides a way to assess these in turn—meta-testing, if you will. It changes your code and reports if your tests still pass.

When you refactor code, you make changes and check the tests still pass. This is a good thing. However, if you randomly change a few plus signs to minus signs, greater than signs to less than signs, change constants, slap in a `break` or `continue`, you would expect some tests to fail or even crash. If they all still pass, that's a bad thing! If you can change the code in malicious ways without any tests failing, are you sure you covered all the use cases? A coverage tool might tell you if the code is called by a test, but if you can change the code, and the test still passes, you may have missed something.

First, you need a test suite you can run automatically. Mutation testing then makes *mutants* by changing the code. If the tests then fail or crash, that mutant is killed. If the tests pass, you have a problem; the aim is to kill all mutants. Mutants can be the simple things (e.g., symbolic changes to operators and constants) and more complicated things (e.g., changing the code path through loops). More advanced mutation tests suites use the abstract syntax tree of your code, allowing even sneakier changes to happen.

Mutation testing packages exist for many languages. There are several for Python, including Cosmic Ray.[4] Austin Bingham has talked about it frequently if you want to explore further.[5]

You point the mutation tester at the genetic algorithm module, ga, tell it where your tests are, . for the current directory, and tell it not to mutate the test code: Boom>cosmic-ray run ga . --exclude-modules=test* It produces a list of the changes that were made and the type, stating if they survive or not. For example:

```
Outcome.SURVIVED -> NumberReplacer(target=0) @ ga.py:9
```

---

4. github.com/sixty-north/cosmic-ray
5. www.youtube.com/watch?v=jwB3Nn4hR1o

This means a mutant that replaced a number with 0 on line 9 of ga.py survived and the tests still passed. This is in random_tries:

```
theta = random.uniform(0.1, math.pi)
```

This mutation will create some initial pairs with an angle of zero. They will get weeded out, but the unit tests with this book don't explicitly test the angle is never zero. In fact, a few mutants survived. This is not a problem but does suggest more tests to add to the code. As it stands, an invalid angle or velocity is never generated, but this is not explicitly tested. The mutation testing made this clear, so it helped find more tests to add. You might find something similar on a code base you are currently working with. Try it out.

Other GA Variations

A variant of genetic algorithms takes randomly generated abstract syntax trees or expression trees and performs mutations and crossover, guided by a fitness function. This is known as genetic programming (GP) and generates programs to solve problems. Mutation testers using abstract syntax trees are a step in this direction. A GA had a fixed-length solution to find. Making a tree is more complicated, but a GP still uses crossover and mutation to create solutions. Trees let you explore other problems, ranging from building decision trees to creating computer programs.

Fuzzers are another way to assess code. They each generate random inputs to try to break your code. The llvm fuzzer generates variants or mutations to increase code coverage.[6] It can find the Heartbleed OpenSSL bug very quickly.[7] Some fuzzers use ideas from genetic algorithms to seek out problems rather than relying on purely random inputs. Genetic algorithms and their variants are powerful!

Over to You

You now have a genetic algorithm to solve a specific problem. You can change the fitness function to solve other problems. You can play around with the parameters—the epochs, the number of items in a population, etc. Let's take a final look at what happened and how you can tweak your solution.

Does your cannon tend to fire balls to the right or left? How often do you get some going one way and some going the other way? Finding more than one solution can be a challenge, so it is covered more in Chapter 10, *Optimize!*

---

6. llvm.org/docs/LibFuzzer.html
7. http://heartbleed.com/

Find the Best, on page 187. How can you encourage solutions to prefer one side over the other? Giving solutions going to your preferred side a bonus in the fitness function works. Or equally, anything going to the other side can drop a few points. Can you encourage the last generation to have solutions going over either side, rather than settling on one?

If you figure out a couple of solutions by hand, one going to the right, and one going left, you can add these to the initial population. This is called *seeding*. If you know a potential solution or two for a problem, feed these seeds to the GA in the setup, or any algorithm that starts with something random. They do not need to be perfect solutions, just something close. This often speeds things up. There is no guarantee a tournament or roulette spin will select these seeds to form the next generation. When new items are bred, your code can save the best few or some specific solutions by adding them to the new population, as though they live for longer.

There are still more options. You can either breed fewer children over time or let the population size vary. If your solutions are not improving, try adding more children; whereas if you want to narrow down on one solution and things are improving, try fewer. How many do you keep alive? The best one? Of course, if you want some to go right and some to go left, you need at least two. You can keep all the solutions that work, or kill the worst few. So many options! Try some out. See what happens.

There are many other things to consider. Try varying the parameters. For example, change the size of the bag. Change the number of solutions in each generation. Change the mutation rate. With the metrics for each run, you can now plot graphs of the results. If you have fewer items in a population, do you need more epochs? Can you end up with all your solutions working?

In this chapter, you covered some core concepts in machine learning and considered closed-form solutions and brute force as alternatives. You reviewed some fitness functions. You learned about the idea of trying something, then looping for a while to get better results. Nature-inspired crossover and mutation are specific to genetic algorithms, but natural evolution or behavior underpins many algorithms. You used probabilistic methods in the mutation function. You will see further probabilistic approaches over the rest of this book. You also touched on genetic programming and mutation testing.

You used some physics equations to model movement in this chapter. The next chapter, Chapter 4, *Swarm! Build a Nature-Inspired Swarm*, on page 57 introduces a model-free algorithm. This machine learning approach iteratively improves a solution using a particle swarm optimization.

Swarm algorithms have a different feel, but they are similar to genetic algorithms. They offer a solution when brute-force fails, they are both *biologically inspired*, and they each offer a population of potential solutions to a problem. Your genetic algorithm worked by killing off unfit solutions. In contrast, a particle swarm emphasizes co-operative behavior. No particles will die off in the next chapter.

CHAPTER 4

Swarm! Build a Nature-Inspired Swarm

In the previous chapter, the paper bag contained a cannon, and you used a nature-inspired genetic algorithm to fire cannonballs out of a paper bag. In this chapter, the bag contains particles which can move independently, follow one another, or swarm as a group. A single particle moving at random might end up outside a paper bag. Several particles moving in a cluster are likely to stay in one clump but might escape if they are each allowed to have some random movement. Several particles following one another but guided via a fitness function will escape the paper bag. These particles will use a *particle swarm optimization* (PSO) to discover how to escape. By sharing information, they gradually move together to achieve this. Some of your cannonballs didn't make it out. Here's a way to avoid this.

The *k-nearest neighbor* (KNN) clustering algorithm can be used to group particles together. This algorithm is handy for finding groups within datasets. It can be used to spot anomalies, find recommendations for music or shopping, or even group machine learning algorithms to spot similar and/or different approaches. The nearest points are found using a *distance* measure. There are several different measures, and the choice can impact the performance of the algorithm. We'll repurpose it to make the particles move, tending to swarm together. By letting each make a small random move too, some might get out of a paper bag. KNN doesn't help the particles escape the paper bag, so you need a better idea. If particles continue tending to follow each other, but also keep track of directions to the better places, they can start to learn. Using a fitness function to decide which trajectory is good does a much better job. You will then see particles following each other, eventually swarming out of the paper bag.

Particle swarm optimizations are fun, simple to code, and applicable to many problems. Learning them increases your knowledge of machine learning, giving you a different slant on nature-inspired algorithms. They are a part of machine learning known as *swarm intelligence*, and learning one will give you a solid basis to try others. PSOs have a similar feel to genetic algorithms, starting with a few random attempts to solve a problem, and then iteratively improving over time. A PSO can be used for many numerical problems—including ones with many *dimensions*. The paper bag in this chapter is two-dimensional but can be extended to three dimensions. Some problems need to find four, or more numbers.

PSO can also find the ideal settings for control systems in dynamic environments, such as liquid levels in tank and for real-time asset allocation.[1,2]There are many application areas.

Some problems can be solved mathematically. However, some problems are *intractable*, so trying something random and incrementally improving is a good alternative. Some machine learning methods use different approaches, for example *kernel* methods. These use mathematics to find *features* to describe patterns in data. However, the random search heuristic is quite pervasive and avoids some of the more difficult mathematics.

Both KNN and PSO can be used to solve problems which *artificial neural networks* (ANN) can solve, but they can both cope with a *dynamic environment*. An ANN is often trained on a static dataset, so never learns new patterns if the environment changes. To analyze a large code base—something which changes if it's being actively worked on—you can find clusters of source files with similar properties (lots of bug reports, people working on it, etc.) and maybe spot which ones are tending to cause the most trouble. Adam Tornhill wrote an excellent book *Your Code as a Crime Scene [Tor15]* exploring code-base changes. He presents various ways to see clusters and patterns, without going into formal machine learning.

Your Mission: Crowd Control

The genetic algorithm helped cannonballs escape a paper bag, starting with random attempts and improving attempts using crossover and mutation. A PSO is another heuristic random search using particles. This time, you'll draw on the HTML5 canvas using JavaScript.

You'll start by moving a single particle around the canvas. Once you know how to move a single particle and draw on canvas, you will consider how to get several

---

1. dl.acm.org/citation.cfm?id=2338857

2. scholarworks.iupui.edu/handle/1805/7930

particles to follow their neighbors using the k-nearest neighbors' algorithm. Finally, you adapt this into a particle swarm optimization algorithm to herd the particles out of the paper bag. The swarm needs to share information as they move, and you'll see how to accomplish this in *A Particle Swarm*, on page 65.

A Single Particle

To get things started, you need an HTML canvas and a button that creates a single particle in a paper bag:

Swarm/paperbag.html
```html
<html>
  <head>
    <title>Particles escaping a paper bag</title>
    <script type="text/javascript" src="paperbag.js"></script>
  </head>

  <body>
    <h1>Can we program our way out of a paper bag?</h1>
    <h2>using one particle moving at random</h2>

    <canvas id="myCanvas" width="600" height="600">
      Your browser does not support the canvas element.
    </canvas>

    <br>
    <p id="demo">Let's try</p>
    <button type="button" id="Go" onclick="init()">Start</button>

  </body>
</html>
```

HTML5 Canvas

Not all browsers support the canvas; be sure to display a default message when that happens. If you use Firefox, Chrome, or a recent version of IE, the canvas should work.

The button onclick event calls an init function in the JavaScript code, shown below. This materializes a particle in the middle of a bag. You use setInterval to repeatedly call the update function to move your particle. Clearing the interval for the given id stops the call, so you can make the button halt the movement using the id:

Swarm/paperbag.js
```javascript
var id = 0;

function Particle(x, y) {
  this.x = x;
  this.y = y;
}
```

```
function init() {
  var c=document.getElementById("myCanvas");
  var particle = new Particle(c.width/2, c.height/2);

  if (id === 0) {
    document.getElementById("Go").innerHTML="Stop";
    id = setInterval(function() {
      update(particle);
      },
      100);
  }
  else {
    clearInterval(id);
    document.getElementById("Go").innerHTML="Start";
    document.getElementById("demo").innerHTML="Success";
    id = 0;
  }
}
```

Making the 100 millisecond interval smaller causes quicker movement. Alternatively, you can call your function once using setTimeout and decide whether or not to call it again afterwards.[3]

Write an update function to move the particle and redraw the canvas as seen in the following code (make your draw function tell you whether or not the particle escaped; if it escaped, reset for another go by calling init again):

Swarm/paperbag.js
```
function update(particle) {
  move(particle);

  if (!draw(particle)) {
    init();
  }
}
```

To move the particle along the horizontal or vertical axis, pick a random step:

Swarm/paperbag.js
```
function move(particle) {
  particle.x += 50 * (Math.random() - 0.5);
  particle.y += 50 * (Math.random() - 0.5);
}
```

The built-in Math.random function returns a number between 0 and 1. Without scaling, your move only goes right or down, by less than a pixel. Subtracting a half gives you a number between -0.5 and 0.5, letting you move in any direction. Multiply your random numbers by 50 to get something between -25.0 to 25.0,

---

3. stackoverflow.com/questions/729921/settimeout-or-setinterval

a reasonable step around the 600 by 600 canvas you defined in the HTML on page 59. Now the steps are much larger than a single pixel. Try out different scales if you'd like.

Now you can draw the bag, and particle and see if the particle escaped:

Swarm/paperbag.js
```
function draw(particle) {
    var c=document.getElementById("myCanvas");
    var ctx=c.getContext("2d");

    ctx.clearRect(0, 0, c.width, c.height);        //clear
    ctx.fillStyle="#E0B044";
    bag_left = c.width/3;
    bag_top = c.height/3;
    ctx.fillRect(bag_left, bag_top, c.width/3, c.height/3);     //draw bag

    ctx.beginPath();
    ctx.rect(particle.x, particle.y, 4, 4);
    ctx.strokeStyle="black";
    ctx.stroke();                                  //draw particle

    return in_bag(particle,
        bag_left, bag_left+c.width/3,
        bag_top, bag_top+c.height/3);
}
```

Get the canvas from your document and from this get a *context object* (ctx). Use ctx to clear the whole canvas, then draw a rectangle for the bag and one for your single particle.

There are a couple of ways to draw rectangles. You can use a filled-in rectangle (filledRect) to represent the paper bag. Fill a third of the canvas (height and width) using fillStyle to set the color. Use an outline of a rectangle (rect) for your particle. Choose a strokeStyle to set the edge color.

Now check if a particle is inside the bag. The sketch on page 62 shows a paper bag and a particle against some axes. Notice the y values are zero at the top and increase as they go down. Any particle in between the top and bottom and between the left and right is inside your paper bag. Any particle going beyond one of these edges escaped.

Check if the particle is within these edges in your in_bag function:

Swarm/paperbag.js
```
function in_bag(particle, left, right, top, bottom) {
    return (particle.x > left) && (particle.x < right)
            && (particle.y > top) // smaller is higher
            && (particle.y < bottom);
}
```

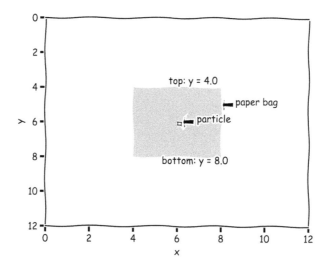

Congratulations! You now have a paper bag and a particle that might escape. The particle zig-zags around and stops if and when it gets out of the bag. In theory, it may never escape; or you might have to wait for ages until it does. Don't forget you can change the interval in setInterval to alter the speed of the movement.

Multiple Particles

Now you can add more particles in the middle of the bag when the button is clicked. The first will continue to move around randomly. Each new particle will also move randomly but will nudge toward its nearest neighbors. Although this section won't have code, you will learn about the k nearest neighbors (KNN) algorithm—*Follow Your Neighbor*, on page 70 walks through the code. To find what's nearest you need to decide a way to measure the distance between particles. The distance measure can have a dramatic effect on an algorithm, which is considered next on *Finding Clusters*, on page 64. Then there's a brief overview of how PSO works in *A Particle Swarm*, on page 65, and after that you are ready for the details of both algorithms.

This KNN algorithm finds k items which are nearest. Many machine learning algorithms need a variety of parameters and choices like distance measures. In this case, you need to decide two: how many neighbors (a value for k), and how to define the nearest (Distance as the crow flies? Another metric? Cost of petrol? Etc.).

Instead of finding several neighbors, you can make a particle step toward its nearest neighbor—using Pythagoras' Theorem to find the closest. The best number of neighbors to use varies from problem to problem, though some-where between 3 and 10 is often a good place to start.[4] The next section describes how to find distances; skip on to *Finding Clusters*, on page 64 if you already know what a *Euclidean distance* is. Armed with the distances to all the other particles, sort these and take the first few to get the nearest neighbors. A particle then steps toward their mid-point, thereby stepping toward the *nearest* neighbors. So, how do you find the distances between points?

Finding distances

For a right-angled triangle use Pythagoras' Theorem, summing the square of the sides at right angles, and square root this to get the longest side or so-called hypotenuse. The figure shows an example (if you can't remember how or why this works, there are some great online resources, giving many different proofs[5]):

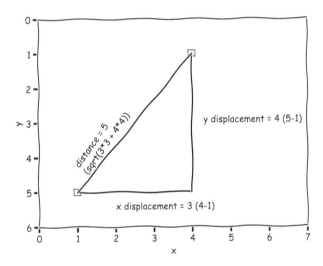

This is called the Euclidean distance. The distance between two points (x_1, y_1), (x_2, y_2) is therefore

$$\sqrt{(x_1 - x_2)^2 + (y_1 - y_2)^2}$$

This distance measures a straight line between the points. For other types of space—curved, for example—you need a different calculation, as the three

4.   http://www.saedsayad.com/k_nearest_neighbors.htm
5.   www.cut-the-knot.org/pythagoras/index.shtml

triangles in the next picture suggest (one might be on a sphere; one might be in curved space-time, as relativity requires):

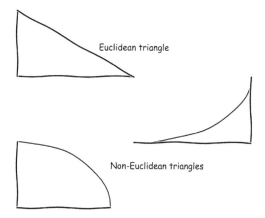

In general, a distance is a *metric*—a function, f with four properties:

- It is never negative
- If the metric between points is zero they are the same point
- The metric from x to y is the same as the metric from y to x
- The metric f(a,c) is never greater than the sum of f(a,b) and f(b,c); going via somewhere else is never shorter!

Many functions have these properties. The metric used can have a big effect on the clusters or neighbors your KNN algorithm finds.

Finding Clusters

This algorithm is a general-purpose clustering algorithm—it will group items into similar groups or clusters. The outcome depends on how many clusters you ask for, how you measure the distance, and how you *encode* your data.

Your particles have two *dimensions*: an x and y coordinate. Adding a z coordinate extends them to three dimensions, and adding yet more makes the algorithm applicable to arbitrary data with many dimensions. Imagine data on people. What dimensions or features might this have? Height, weight, age, gender, favorite band, favorite book, address? The feature choice has an influence on the clusters formed, and most data has lots of dimensions.

The units matter too. Some of the numeric data will cover very different ranges; 180 cm is possible, but 180 years old is unlikely. Some distance functions or metrics need values to be scaled to the same range to work well; otherwise, the bigger numbers are dominant. You can normalize data, ensuring all values end up on the same scale or a standard statistical model by shifting and scaling values. This is part of data *pre-processing*, which is a huge topic. The Python scikit-learn machine learning library has a great list of scaling techniques if you want to know more.[6]

Some of the features are not numeric—they give categories instead. An address can be transformed into GPS coordinates or kept as a string. These representations need different distances measures. Machine learning uses many different distance measures or metrics!

Let's consider an example. Responses to a survey about favorite books may have answers in the form of like versus do not like. Each response can be represented as a list [0, 1, 0, 0, ..], [1, 0, 1, 1, ..], [0, 0, 1, 1, ..]. Trying to plot this on a graph to find groups by eye needs some thought! Instead, you can use the KNN algorithm, using the Euclidean distance to find neighbors. Since each data point is a one or zero, you can skip some of the calculation by counting how many bits differ. This is called the *Hamming distance*. Responses like [0, 1, 0, 0] and [1, 0, 1, 1] completely disagree, so have Hamming distance 4. In contrast, [1, 0, 1, 1] and [0, 0, 1, 1] agree on three of the four bits, so have a distance of 1. In some sense, they are closer, or more similar.

Clusters give you a way to analyze data. Some items group together while others stand out on their own as *outliers*. Clustering is a type of *unsupervised* machine learning; it doesn't predict anything specific, or build a model, or solve a problem. It does, however, help you learn about your data.

A Particle Swarm

The KNN algorithm can be re-purposed to move particles around. How likely is it that particles following each other will escape the bag? Not very. If you add some random movement too, one or two may escape, but they will tend to meander near each other, which is a disaster. If you add some guidance, rather than small random nudges, things change.

If your particles track the best place so far (maybe the highest), and head that way, they are more likely to get out of the bag. To do this, you need a fitness function, as you had before on page 42, to find the best. Making higher y positions better encourages upward motion, so your particles tend to travel up and out of the top of the bag. What if you only have one particle? If it only goes to the best place so far, it will never move! Some random exploration will overcome this, just as the random move with KNN helped. Adding several more particles to explore more places helps even more. If all the particles go

---

6.   http://scikit-learn.org/stable/modules/preprocessing.html

to the same place, there won't be much to see either. By combining the best position overall and each personal best, the particles will make a particle swarm optimization (PSO).

This PSO algorithm is a part of machine learning known as swarm intelligence. There are many variations of swarm algorithms, and you will see other algorithms later in this book. They involve *agents*: particles, bees, ants, and more besides, exploring nearby and sharing information to build a big picture globally. Combining the global and local information gives you a swarm moving toward an *optimal* point.

Like GA, you start with something random and improve. Unlike GA, you don't need crossover and mutation and don't need the physics model of ballistics. You do, however, need to find a way to measure the best and think of a way to combine the local and global information. Different swarm algorithms take different approaches.

You're now ready for the details; you covered a lot of new terms so far. Take a breath. Sit back and watch your one particle doing its random walk. The next section creates a swarm of particles following each other, and then a much better swarm swooping out of the paper bag.

How to Form a Swarm

You built the single particle code already. Drawing several particles is similar. They can even move randomly but don't improve over time if that is all they do.

First, get the particles following their neighbors by finding these and making a move function to nudge each particle toward these. Unsurprisingly, these tend to gang together and take a long time to escape the paper bag. By making a different move function, the particles swarm out of your paper bag. The individual movement of the swarm particles will combine momentum, picked at random, to begin with. Over time, the movement needs to include movement toward a personal best and the global best to share information through the swarm. The standard PSO calculates a weighted sum of the current velocity, distance to the personal best and swarm's best place so far to make a move. This section will show you how.

The single particle moved at intervals. Several particles can move one at a time, or in a batch. You'll try both methods—one at a time when you follow the neighbors and all together for the swarm. Many machine learning algorithms have both flavors, and it is not always obvious which to use, without trying out both.

Follow Your Neighbor

To follow their neigbors, move each particle by a small random nudge and a step toward the mid-point of its nearest neighbors. You need to decide how many particles to have and how many neighbors to follow. If you make a new particle on a button click, you can vary the number of particles easily. Try five neighbors to begin with. Watch out for fewer than five particles in total—use all the particles if you don't have enough.

The code is similar to *A Single Particle*, on page 59, but the move function nudges particles toward their k nearest neighbors:

```
On click:
  Kick off a particle
  setInterval(update(particle), interval(150))

On update:
  move(particle).randomly()
  neighbors = find_knn(5)
  move(particle).mid_point(neighbors)
  draw()
```

When all the particles update together the algorithm uses *synchronous* or *batch update*. Update each particle by itself to make an *asynchronous* algorithm. To change between these, loop through all of the particles in the update instead of only updating one.

The random nudge keeps the particles moving. Without this, they all bunch up and may never make it out of your paper bag. This KNN algorithm has no sense of purpose, so the particles tend to amble around for quite a long time. They will eventually escape your paper bag if you are patient.

Follow the Best

To encourage the particles out of the paper bag, drop the nearest neighbor part of the code and bake in some intelligence or purpose. If individual particles remember their best spot so far and all the particles compare notes on the overall best spot, you can make them move toward a combination of these. There are different ways to determine the best position. Taking the highest position makes particles tend to go up. Finding a position nearest any edge of the paper bag allows particles to go through the sides of the bag. If you have some near the right edge and some near the left, you are in danger of heading toward the center. Using the highest position as a fitness function is bound to work.

Armed with a fitness function to find the best positions, your algorithm looks like this:

```
Choose n
Initialize n particles randomly
For a while:
  Update best global position (for synchronous)
  Draw particles at current positions
  Move particles (for asynchronous: update global best here instead)
  Update personal best position for each particle
```

Stop your loop when all of the particles are out of the paper bag, or after a few iterations. You decide.

Each particle has an (x, y) position and a velocity. The velocity gives the speed in steps per time period and a direction of movement. Adding the current horizontal velocity v to the current x position gives the next x position:

$$x_{t+1} = x_t + v_{x,t+1}$$

Find the new y position using y in place of x. For three dimensions, use a z coordinate as well, and update similarly. To find the best of many parameters or features, use one dimension per feature. Using your paper bag and moving particles on the canvas gives a visualization of the learning in the algorithm. Higher dimensions are harder to visualize!

By using suitable velocity, you get a swarm. Each particle has its very own velocity, initialized at random. Over time, it tends toward a sum of local and global information. A particle's personal best (p) is the local information, and the swarm's global best (g) is the global best. Using the y value is a straight-forward way to find the best. Higher up is better.

The standard way to update the velocity v adds pre-chosen fractions or weights of the distances to the personal best and global best to a fraction of the current velocity. This gives a particle momentum along a trajectory, making it veer toward somewhere between its best spot so far and the overall, or global, best spot. Using weight or fraction w for the current velocity, c_1 for distance to the personal best, and c_2 for distance to the global best gives:

$$v_{x,t+1} = w * v_t + c_1 * (p_t - x_t) + c_2 * (g_t - x_t)$$

To find the new y velocity, you guessed it, use y in place of x.

If w is zero, the particle forgets what happened, since the last velocity encoded local and global best spots. If the particles start in the same place, they all stay still because the best is the same as the worst. Everything is the same.

Forever! If particles start in different places, they will move, but they will not swarm. So, try something greater than zero for w. Experiment with the other weights. A larger value for c_1 prefers the personal bests, so the particles may explore more. If c_2 is larger, they will stay closer together.

The initial random velocity gives some variety, but not much. Using a small *stochastic* (random) weight along with the parameters gives more variety, and you form a swarm. To do this, scale c_1 and c_2 by random numbers r_1 and r_2. The particles can then explore more, and you get a proper swarm:

$$v_{t+1} = w * v_t + r_1 * c_1 * (p_t - x_t) + r_2 * c_2 * (g_t - x_t)$$

Random numbers between -5 and 5 work well for a 500 by 500 bag, since they make relatively small moves. Larger jumps make the movement look jerky. You can decrease the upper limit of 5 over time to make the swarm group more tightly. For another bag size, you might want to change the maximum move. Each problem needs slightly different parameters. Play around and see what seems to work. You can even make w, c1, and c2 vary over time, or with the quality of the results.

Parameter Choice

If you tune your parameters to solve one problem well, you may find you need to re-tune them to solve another problem. This suggests the machine isn't learning anything! This is a big problem for machine learning algorithms trying to model data; they often *overfit* a training set and fail badly on new data.

You saw that the global best position could be updated synchronously or asynchronously in the algorithm code on page 68. To code the asynchronous version call the update function after each particle moves. To code the synchronous version call the update function after all the particles move. Several variants of the algorithm exist, and you might be able to dream up your own.

Let's Make a Swarm

Now that you know what you're going to do, and you have an idea of how to do it, you're ready to dive in and write the code. For both algorithms, KNN and PSO, you can reuse the HTML from the code on page 59 to create a canvas and have a button calling an init function. Both algorithms need Particles and a move function. After each move, draw the current positions on your canvas to see what's happening.

Follow Your Neighbor

To make your particles follow their neighbors, create an array of particles. Add one each time this button is clicked. Use setInterval to drive a particle's movement, and save the interval's id in the class so you can make it stop. You also need to store its index into the array and the current x and y position. Start each particle at a random position inside the bag, ready to move:

Swarm/src/knn.js
```
var bag_size = 600;
var width = 4;
var left = 75;
var right = left + bag_size;
var up = 25;
var down = up + bag_size;

function Particle(x, y, id, index) {
  this.x = x;
  this.y = y;
  this.id = id;
  this.index = index;
}

var particles = [];

function init() {
  var x = left + 0.5 * bag_size + Math.random();
  var y = up + 0.5 * bag_size + Math.random();
  var index = particles.length;
  id = setInterval(function() {
        update(index);
      },
      150);
  var particle = new Particle(x, y, id, index);
  particles.push(particle);
  document.getElementById("demo").innerHTML="Added new particle " + index;
}
```

The array of particles move about randomly, just as the single particle did, but nudging toward their nearest neighbors with each step in the update function. Notice it is called for a specific particle; each particle updates asynchronously, rather than them all moving in lock-step. Once a particle is out of the paper bag, clear the interval, so it stops:

Swarm/src/knn.js
```
function update(index) {
  var particle = particles[index];
  move(particle);
  draw();
```

```
  if (!in_bag(particle, left, right, up, down)) {
    document.getElementById("demo").innerHTML="Success for particle " + index;
    clearInterval(particle.id);
  }
}
```

When a particle moves, it has to find its nearest neighbors. You want the distance between a given particle, identified by its index in the array, and all the other particles. Pythagoras' Theorem tells you the Euclidean distance between particles. You can use other distance functions here if you wish. If you pair the distance with the index into the array and sort from nearest (smallest distance) to furthest, taking the top k will give the indices of the neighbors you need.

Swarm/src/knn.js
```
function distance_index(distance, index) {
  this.distance = distance;
  this.index = index;
}

function euclidean_distance(item, neighbor) {
  return Math.sqrt(Math.pow(item.x - neighbor.x, 2)
                 + Math.pow(item.y - neighbor.y, 2));
}

function knn(items, index, k) {
  var results =[];
  var item = items[index];
  for (var i = 0; i < items.length; i++) {
    if (i !== index) {
      var neighbor = items[i];
      var distance = euclidean_distance(item, neighbor);
      results.push( new distance_index(distance, i) );
    }
  }
  results.sort( function(a,b) { return a.distance - b.distance; } );
  var top_k = Math.min(k, results.length);
  return results.slice(0, top_k);
}
```

Use this in your move function. Make a random move as before, but much smaller, giving more weight to the neighbors' positions:

Swarm/src/knn.js
```
function move(particle) {
  //first a small random move as before
  //with 5 instead of 50 to force neighbors to dominate
  particle.x += 5 * (Math.random() - 0.5);
  particle.y += 5 * (Math.random() - 0.5);

  var k = Math.min(5, particles.length - 1);//experiment at will
```

```
var items = knn(particles, particle.index, k);
var x_step = nudge(items, particles, "x");
particle.x += (x_step - particle.x)
                    * (Math.random() - 0.5);
var y_step = nudge(items, particles, "y");
particle.y += (y_step - particle.y)
                    * (Math.random() - 0.5);
}
```

Find the mid-point of the neighbors in the x or y axis in your nudge function. Then you can make the current particle step left (or right) a bit and up (or down) a bit. The "bit" is controlled by randomness. Average the x (or y) coordinates to find a point to aim for:

Swarm/src/knn.js
```
function nudge(neighbors, positions, property) {
  if (neighbors.length === 0)
    return 0;
  var sum = neighbors.reduce(function(sum, item) {
    return sum + positions[item.index][property];
  }, 0);
  return sum / neighbors.length;
}
```

Feel free to fiddle around with the number of neighbors as well as the distance algorithm. If the random first step is much larger than the step toward the neighbors, the particles tend to move relatively independently. Your last move (code on page 60) scaled up the step by 50; this uses 5. Play around with it. Can you find a critical point? Does it vary with the number of particles?

Follow the Best

Now you can code your PSO. Again, you need an HTML file to drive this. It needs a canvas and a button calling the init function to set up an array of particles:

Swarm/src/pso.js
```
var id = 0;

function makeParticles(number, width, height) {
  var particles = [];
  var i;
  for (i = 0; i < number; ++i) {
    x = getRandomInt(0.1*width, 0.9*width);
    y = height/2.0;
    var velocity = { x:getRandomInt(-5, 5), y:getRandomInt(0, 5)};
    particles.push ( { x: x,
                       y: y,
                       best: {x:x, y:y},
                       velocity: velocity } );
  }
```

```
      return particles;
}
function init() {
  if (id === 0) {
    var canvas = document.getElementById('myCanvas');
    document.getElementById("Go").innerHTML="stop";
    particles = makeParticles(20, canvas.width, canvas.height);
    var epoch = 0;
    draw(particles, epoch);
    var bestGlobal = particles[0]; //or whatever... pso will update this
    id = setTimeout(function () {
          pso(particles, epoch, bestGlobal, canvas.height, canvas.width);
        },
        150);
  }
  else {
    clearInterval(id);
    id = 0;
    var canvas = document.getElementById('myCanvas');
    document.getElementById("Go").innerHTML="go";
  }
}
```

The particles have an x and y value again. They also have a personal best and a velocity. The loop now uses a setTimeout to pass in the parameters. You call pso again on a timer, for a while.

The particles start halfway up (or down) the canvas, at a random width. You can change this—if they all start in the middle the swarm movement changes. The velocity also impacts the movement. If you use { x: 0, y: 0 } and everything starts at the same height they will move from side to side. Play around with these values. See what happens.

Move the particles in your pso function, draw them at their new positions and then update the best:

Swarm/src/pso.js
```
function pso(particles, epoch, bestGlobal, height, width) {
  epoch = epoch + 1;
  var inertiaWeight = 0.9;
  var personalWeight = 0.5;
  var swarmWeight = 0.5;
  var particle_size = 4;
  move(particles,
    inertiaWeight,
    personalWeight,
    swarmWeight,
    height - particle_size,
    width - particle_size,
```

```
    bestGlobal);
  draw(particles, epoch, particle_size);
  bestGlobal = updateBest(particles, bestGlobal);
  if (epoch < 40) {
    id = setTimeout(function () {
      pso(particles, epoch, bestGlobal, height, width);
    }, 150);
  }
}
```

Move using a combination of current position, personal best, and global best, making sure you don't go beyond the edges of the canvas. The combination needs several magic numbers or parameters. Play with them and see what difference it makes:

Swarm/src/pso.js
```
function move_in_range(velocity, max, item, property) {
  var value = item[property] + velocity;
  if (value < 0) {
    item[property] = 0;
  }
  else if (value > max) {
    item[property] = max;
  }
  else {
    item[property] = value;
    item.velocity[property] = velocity;
  }
}

function move(particles, w, c1, c2, height, width, bestGlobal) {
  var r1;
  var r2;
  var vy;
  var vy;
  particles.forEach(function(current) {
    r1 = getRandomInt(0, 5);
    r2 = getRandomInt(0, 5);
    vy = (w * current.velocity.y)
        + (c1 * r1 * (current.best.y - current.y))
        + (c2 * r2 * (bestGlobal.y - current.y));
    vx = (w * current.velocity.x)
        + (c1 * r1 * (current.best.x - current.x))
        + (c2 * r2 * (bestGlobal.x - current.x));
    move_in_range(vy, height, current, "y");
    move_in_range(vx, width, current, "x");
  });
}
```

This uses a getRandomInt utility function to generate a random number between min and max:

Swarm/src/pso.js

```
function getRandomInt(min, max) {
  return Math.floor(Math.random() * (max - min + 1)) + min;
}
```

Drawing the particles is straightforward, though notice this time you subtract the particle's y value from the canvas. This means 0 is at the bottom and the largest number is at the top. This seems more natural to some people, but if you are happy with 0 at the top, that's fine:

Swarm/src/pso.js

```
function draw(particles, epoch, particle_size) {
  var canvas = document.getElementById('myCanvas');
  if (canvas.getContext) {
    var ctx = canvas.getContext("2d");
    ctx.clearRect(0, 0, canvas.width, canvas.height);
    ctx.fillStyle = "rgb(180, 120, 60)";
    ctx.fillRect (2.5*particle_size, 2.5*particle_size,
                  canvas.width - 5*particle_size,
                  canvas.height - 5*particle_size);

    var result = document.getElementById("demo");
    result.innerHTML =  epoch;

    particles.forEach( function(particle) {
      ctx.fillStyle = "rgb(0,0,0)"; //another way to spell "black"
      ctx.fillRect (particle.x,
              canvas.height - particle.y - particle_size/2,
              particle_size, //width and height of particle - anything small
              particle_size);
    });
  }
}
```

Find the global and personal best in updateBest. Remember, higher is better. If you have 0 at the bottom of the canvas, then bigger numbers are better. If you stuck with 0 at the top, you just need to flip the greater than sign in the best function to a less than since with 0 at the top smaller numbers are better:

Swarm/src/pso.js

```
function best(first, second) {
  if (first.y > second.y) {
    return first;
  }
  return second;
}
```

```
function best(first, second) {
  if (first.y > second.y) {
    return first;
  }
  return second;
}
```

You don't need to go up—you can see which was nearest an edge. Or use anything else you can dream up.

That's it! A random start and iterative improvement in a loop. You made a nature-inspired swarm using the PSO algorithm. Time to assess what happened.

Did It Work?

As expected, the neighbors tend to follow one another for a while, starting and staying near the initial position, as shown below on the left. Eventually, some escape, as shown in the picture on the right:

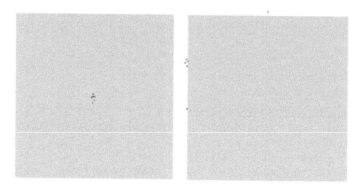

How many escape depends on the ratio of random movement to velocity toward the neighbors, and the number of neighbors you used. Some tend to escape out of different sides as shown in the picture. As one makes a break for it, it moves away from its neighbors. The next to escape appears to move away from the last escapee. Since the purpose of this algorithm was to learn about clustering, it's OK if something different happens every time; this unsupervised algorithm is about finding clusters or groups. You re-purposed it. The rate at which you clicked your button to add particles has an effect too.

Now to the PSO. Remember when you made your particles, you started with a random position somewhere in the middle of the bag? You set the x and y positions using these lines of code:

```
x = getRandomInt(0.1*width, 0.9*width);
y = height/2.0;
```

They form a line and then move, quite quickly, either to the right or left, swarming up as they go. This should be no surprise—the fitness function picked the higher up of any two particles. The next pictures show particles starting along a middle line on the left. Over time they head to the left, as the middle picture shows, then swarming up and out when they start in different places, as the rightmost picture shows:

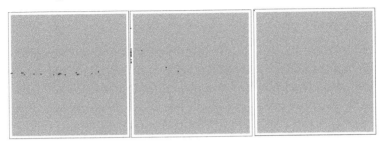

If you start all the particles in the middle, by changing the initial value for x to width/2, they tend to go diagonally as seen in the next picture, rather than heading to one side as you just saw. Try letting them start at a few different heights and see what happens. Play with the number of particles.

There are many different options to try. Can you recall how many there were for the GA? Epochs, generation size, rate of mutation, and more besides.

This time around you had some similar parameters: You used 20 particles for your swarm size. You don't have epochs—in a sense no particles die off—but you stopped the loop at 40. Alternatively, you can wait until everything escaped. You didn't need a mutation rate, but you did need three magic numbers, w, c1, and c2; inertia, personal weighting, and global weighting. If w is 0, and they all start at the same height, there is no vertical movement. So, no inertia makes the particles move less! When c1, the personal best, is much larger, they tend to explore independently. When c2, the global weight, is much larger, they tend to move together. The shape of the swarm varies with the parameters and gives you an idea of how the particles will explore.

Over to You

You covered a lot in this chapter. In later chapters, you will see other swarm intelligence algorithms, which will reinforce some of the ideas here. You also used a clustering algorithm. You discovered finding a suitable metric can be difficult, and the more parameters you have, the more choices you need to make—sometimes this leads to over-fitting or a model that is no use to anyone else.

Enjoy your moving particles and have fun playing around with the parameters. If you try different fitness functions, you can make the swarm swoop around in a variety of different ways. You can go down, left, or right instead. If you get inventive, you can make it circle or spiral. You can even extend the algorithms to three dimensions or more.

You can use PSO for a variety of real-world problems. James McCaffrey provides a walk through in C# on his blog to find the minimum value of a function and suggests a variety of extensions.[7] Chapter 10, *Optimize! Find the Best*, on page 187 explores finding minima in other ways.

There are several other swarm intelligence algorithms, which all have a similar feel: some random setup, a loop, combining something one agent knows and something the whole swarm knows in various proportions. In the next chapter, your agents will be ants. This swarm algorithm will report a path through your paper bag. Up to this point, you have concentrated on making particles get out of the bag. Next time, they will tell you the route they took. This nature-inspired algorithm still uses co-operation—sharing knowledge, but prior attempts get forgotten over time. This approach lends itself to problems that need to find the best path, shortest route, cheapest layout, and many other spatial or state-based problems.

---

7. See http://msdn.microsoft.com/en-us/magazine/hh335067.aspx

CHAPTER 5

Colonize! Discover Pathways

In the previous chapter, you encouraged a swarm of particles to escape your paper bag. This time, imagine a colony of ants exploring in and around a paper bag, searching for food. Your ants can amble around, visiting a few places along the way, but once they find food, they go home. You can persuade the ants to find a path out of the paper bag if you put all the food above the bag. Initially, they will explore inside the paper bag and eventually end up above the bag. They can then feed and go home. Over time, they will learn efficient routes to get back to the food. In fact, when your ants go straight up, they have found the shortest path to take.

In this chapter, you will discover how an *ant colony optimization* (ACO) finds a good path through space. An ACO is similar to a PSO; the agents, now ants, use a quality function to decide the next possible steps. This plays a similar role to the fitness functions you have seen, but instead of maximizing fitness, the ants want to minimize the path length or cost. They also share information by leaving a trail for others to follow. This makes an ACO ideal for finding a route. It may not get the best solution but can give reasonably good answers to extremely tough *combinatorial* problems quickly: anything where you will take forever to try every possible approach. ACO even works if the problem changes dynamically. How do you find the best route across town? It will depend on the current traffic.

A common machine learning puzzle is the *traveling salesman problem* (TSP). The aim is finding the shortest route around several cities, visiting each just once, and ending back where you start. For two cities there's only one distinct route, from one city to the other. As you add extra places, this grows factorially, so checking for the very best will take a long time. You can use an ACO to find possible solutions to this puzzle since it gives a good route to follow quickly, even in many cities.

Marco Dorigio is credited with inventing the algorithm in 1992, using it for the TSP. The original version was called an *ant system*, and the ant colony optimization emerged from this. ACO has two phases, generating solutions and updating the paths using *pheromones*, by addition or evaporation. It also allows so-called *daemon actions*, having no real-world basis I assume. These actions allow you to implement variations easily. Dorigo and Stutzle, authors of the definitive textbook on ACO *Ant Colony Optimization [DS04]*, refer to daemon actions as "centralised actions executed by a daemon possessing global knowledge."

Your Mission: Lay Pheromones

Real-life ants do communicate, admittedly indirectly, via pheromones in their environments. Evaporation doesn't always take place in nature. The ant colony algorithm is inspired by this natural behavior. Many machine learning approaches are nature inspired.

The overall shape of the algorithm itself will be familiar: you start with something random—this time a path walked by fictitious ants—and iteratively improve. Your ants will leave a trail of pheromones as they explore. Over time, the previous pheromones evaporate, and the ants lay down new pheromones as they find new paths. The ants build a path one point at a time. The point can be a physical place or a state. In this chapter, the ants choose between nearby points, using a combination of path length and the pheromone level. Roulette wheel selection is a great way to choose a spot, which you encountered in Chapter 3, *Boom! Create a Genetic Algorithm*, on page 33. Alternatively, you can use a tournament selection or always pick the best. A tournament can cause premature convergence, with the ants settling down on worse paths. Always picking the best might miss better options too. The ants will then start following the best path so far, and might not find even better paths. For this recipe, it doesn't make much difference; however, keep in mind picking slightly worse spots or solutions allows your machine learning algorithm to explore more. If you use tournament selection, you need to decide how many spots compete. The roulette wheel method doesn't need this extra parameter choice, so use that. Furthermore, most ACO uses a proportionate probabilistic selection, like roulette wheels.

Spots with more pheromones are more appealing, so you need to put more pheromones at points along the shorter paths. The reciprocal of the path length works well. You also bake in a heuristic—for example, go to the nearest city for the traveling salesman problem. In this chapter's problem, you will pretend you have lots of ant food above the bag, so the ants are more likely to find it if they go up. A well-chosen heuristic will help your algorithm find better solutions more quickly.

Why the reciprocal? Consider the two ant paths shown in the following diagram, one of which is longer than the other:

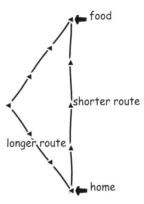

If the ants use the same total amount of pheromone for each trip, spots (or nodes, if you are familiar with graphs) on the shorter path (or edge) will be more intense. The ants leave 1/4 of the total pheromone at each point on the short path, but only 1/6 to each point on the longer path. The shortest path is now more appealing.

When you implement this, you can either update pheromones as you go, once a single ant gets home, usually called a nest, or when they all get home. The ACO tends to call these flavors:

- Online step by step—each ant updates pheromones as it explores
- Online delayed—each ant updates when it gets home
- Offline pheromone—wait until all ants are home

In this chapter, you will wait until all the ants are home. This makes the code slightly simpler. It's not difficult to implement the other versions, but let's start with the simplest approach. You can try out the other versions too. Whichever you use, the shorter path smells more attractive.

Once home, the ants set off again, guided by the pheromones. The ants will tend to move toward the most promising routes. In fact, they will all tend to move toward the same route if that's all you do. Over time, you must decrease the previously laid pheromone levels, so your ants get a chance to explore properly. In time, they'll find better paths, solving your problem.

Using the Pheromones

Your selection method will probabilistically pick one of the next possible spots on a path. You give each spot a score combining the pheromone (τ, pronounced "tow," as in the start of the word tower, and written as "tau" in English letters) and a quality value (η, pronounced "eater" and spelled eta). We'll multiply these together to make a taueta function later. The quality is a heuristic to encourage specific behavior; it may not be needed in general. The canonical combination multiplies powers of both these numbers. Bigger numbers are more likely to win. The exact powers, α and β , are up to you. Try different values to see what happens.

The chance of moving to a given spot or state (i) of all possible states (j) is

$$p(\text{spot}_i) = \frac{\tau_i^{\alpha} \times \eta_i^{\beta}}{\sum_j \tau_j^{\alpha} \times \eta_j^{\beta}}$$

You divide by the total, so the sum of all these is 1, giving you a probability of choosing a spot. 0 never gets chosen; 1 is bound to be chosen. Anything in between may or may not get picked. This only works if you avoid all the values being zero. You can give a minimum pheromone value to each spot to ensure this. Alternatively, you can pick any possible move at random if the sum is zero, to keep it simple. You don't need to divide by the total either. If you find the product:

$$\tau\eta(\text{spot}_i) = \tau_i^{\alpha} \times \eta_i^{\beta}$$

you can still tell which points are better. You will use the reciprocal of the path length to set τ and the y value of a spot for η. This makes high-up spots with more pheromones better. Initially, ants will explore off in various directions. Over time, the ants will tend to make a shorter journey heading out of the paper bag. That's where the food is, and they are hungry.

You know how to assign a metric to each spot and options for choosing between a few possible spots. To evaporate the pheromone, you can use any function to reduce the values without going negative. You nearly have all that's required to ensure some iterative improvement. You just need to decide the possible spots an ant can travel to.

Where Can the Ants Go?

If the ants can walk over an imaginary grid, going in all directions, they have eight possible places they can go. You can see this in the diagram:

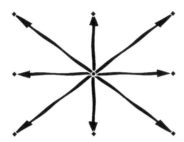

If you stop the ants marching through the sides of your bag, they have fewer options at the edges. Allow them out of the top though! You don't need to stick with these eight options. You can drop down to four cardinal positions; north, east, south, west. You can try hexagonal grids instead. You can probably dream up something completely different.

You don't even need to stick to a grid. In theory, you can let the ants go anywhere they wish, giving you a *continuous ant colony optimization*. Tracking pheromones for every point in space isn't possible, so the continuous extension takes a different approach which won't be covered here.

You now have an overview of how to create an ant colony to solve a problem. The next section considers options for your setup and shows you how to evaporate and update your pheromones so you can build pathways.

How to Create Pathways

In pseudo-code your ACO works like this:

```
for a while
  create paths
  update pheromones
  daemon actions: e.g. display results
```

You will start your ants somewhere along the bottom of the bag, let them walk around grid points, and then return home once they find the food above the top of the bag. The ants can start anywhere along the bottom of the bag, or in the same place. You can either control this by a checkbox from the HTML or hard-code one of these options. The specifics don't make a difference to the overall feel of the algorithm, but the choice can affect how long your ants take to find good paths.

You need to make several decisions:

- How many ants?
- How far apart are the grid points?
- What values to use for the parameters?
- How much to evaporate the pheromones by?

Start with a 5x5 grid—with the top five points above the bag—and about 25 ants. Vary these when you've got this working. You will make a random path for each ant to get the algorithm started. Choose a possible next spot, watching out for the bag edges, and avoiding previously visited spots. Only let an ant revisit a spot if it corners itself.

Now you loop around, getting your ants to learn. First, evaporate the pheromones at some pre-chosen rate ρ, maybe 0.25:

$$\tau_i = (1 - \rho) * \tau_i$$

Let each ant explore until it goes above the top of the bag then teleport it home. When all of the ants are home, update the pheromones. It might seem strange that an ant can somehow put a pheromone at spots on a path once back home. As with most nature-inspired algorithms, some unnatural things might happen. Machine learning and artificial intelligence might be inspired by nature, but it doesn't try to emulate precisely what happens. Imagine ants lay pheromones on the way home if you'd rather.

Use the reciprocal of the path length for the pheromone value to make shorter paths more attractive. You will use each ant's path in the pheromone update. An ACO often uses a constant (Q) to scale the quality of a path. Other letters might be suitable for different problems. You can even use whole words for the variable names in your code! Many machine learning algorithms have a mathematical flavor, so it is useful to be aware of the terse parameter names. The constant Q will depend on your problem—for larger bags use a bigger value, so scale up by the bag height. Calculate the increase in pheromone level (L) like this:

$$Q = 2.0 \times height$$

$$L = Q / length(path)$$

Add this extra pheromone level to each point p on each path:

$$\tau_p = \tau_p + L$$

You now have updated pheromones, and you know a way to let them evaporate. How do you use them to create pathways?

Each ant will build up a path, picking a starting spot first and keeping track of where it went. To decide the next spot, an ant has a choice of up to eight spots. You can prevent an ant from revisiting spots if you want, by looking back at the path so far. Your ant may then have fewer than eight places to try. If you don't stop spots from being revisited your ants may go round in circles for a while. Make a list of spots or points an ant can visit, and use roulette wheel selection as you did before on page 43. Pick a random number between 0 and the total value ($\Sigma \tau \eta$, defined on page 82) of these points, and send your ant to the corresponding point. You're more likely to pick better points but allow some exploration.

You now have all the ingredients you need to create an ACO. You will display the best and worst paths over time, and can report best, worst, and average path length each time. You should see improving paths as your code runs. Time to code it!

Let's March Some Ants

You can code this in JavaScript and display the routes with the HTML canvas. You can reuse the HMTL on page 59. Add an input checkbox to decide if all the ants start in the same place, like this:

```
<input name="middle_start" id="middle_start" type="checkbox">
  Start in middle?
</input>
```

The button calls init so you can set up your ACO, or stop if it is already running. Use the checked property to decide where to start, then begin:

```
Colonise/src/aco_paperbag.js
var id = 0;
var middle_start = false;

function init() {
  if (id === 0) {
    document.getElementById("click_draw").innerHTML="stop";
    var opt = document.getElementById("middle_start");
    if (opt) {
      middle_start = opt.checked;
    }
    begin();
  }
  else {
    stop();
  }
}
```

Random Setup

To begin, make_paths for the ants, just randomly choosing spots, and update the pheromones, then draw the paths. With things set up, you then run your ACO at intervals. This, like the setup, accomplishes three main steps: create paths, update pheromones, and then the daemon action of drawing paths. You'll implement this on page 89. Let's finish the setup first:

Colonise/src/aco_paperbag.js
```
function begin() {
  var iteration = 0;
  var canvas = document.getElementById("ant_canvas");
  var pheromones = [];
  var height = canvas.height / scale;
  var width = (canvas.width-2*edge) / scale;
  var ants = 25;
  var paths = make_paths(height, width, ants);
  update(pheromones, paths, height);
  draw(iteration, paths);
  id = setInterval(function() {
      iteration = aco(iteration, ants, pheromones, height, width);
    },
    100);
}
```

The interval runs every 100 milliseconds, giving the browser a chance to update. Save the id you get from setInterval to stop your aco if the button is clicked again. To stop, clear the interval and set the button text back to "action," ready for another go:

Colonise/src/aco_paperbag.js
```
function stop() {
  clearInterval(id);
  id = 0;
  document.getElementById("click_draw").innerHTML="action";
}
```

You need to make decisions for a few parameters. First, you need a grid size. You could just take a step of one pixel in any direction, giving a large grid. If you scale this to a proportion of the canvas size your ants will complete a journey more quickly. Using a fifth of the canvas height is a good compromise. Set scale to 50.0 for a canvas of height 250, then each x or y step is 50.0 pixels. An ant can then make just five steps up to find food, once it's found the best route.

How many ants? Start with 25 and experiment. For a smaller scale your ants need more steps to get food, so you may need more ants, and you might need to let them explore for longer.

For each ant, start each path somewhere suitable, and add steps until it finds food. You can use an array of {x, y} positions for a path. The food is anywhere with a y value greater than the bag height. Add an edge around the bag, so it doesn't stretch across the whole canvas. Find the start_pos, and add this to your path. Note that the x value might not be a grid point, so you need to floor it. Keep adding the next_point until your ant finds food:

Colonise/src/aco_paperbag.js
```
function start_pos(width) {
  if (middle_start) {
    return { x: Math.floor(width / 2), y: 0 };
  }
  return { x: Math.floor(Math.random() * (width+1)), y: 0 };
}

function random_path(height, width) {
  // Assume we start at the bottom
  //   If we get to the top, we're out so finish
  var path = [];
  var pos = start_pos(width);
  path.push(pos);

  while (pos.y < height) {
    pos = next_pos(width, pos, path);
    path.push(pos);
  }

  return path;
}

function make_paths(height, width, ants) {
  var paths = [];
  var i;
  for (i = 0; i < ants; i += 1) {
    paths.push( random_path(height, width) );
  }
  return paths;
}
```

To find the next_position list the eight possible_positions, and filter out any that let your ant sneak out of the edges. Check your ant's path doesn't contain a point already, but let it revisit spots if your ant runs out of options. Now you can pick a point from the allowed_positions at random to build up the path:

Colonise/src/aco_paperbag.js
```
function possible_positions(width, pos) {
  var possible = [
    {x: pos.x - 1, y: pos.y - 1},
    {x: pos.x,     y: pos.y - 1},
    {x: pos.x + 1, y: pos.y - 1},
    {x: pos.x - 1, y: pos.y},
```

```
      {x: pos.x + 1, y: pos.y},
      {x: pos.x - 1, y: pos.y + 1},
      {x: pos.x,     y: pos.y + 1},
      {x: pos.x + 1, y: pos.y + 1}
  ];

  return possible.filter( function(item) {
    return item.x >= 0 && item.x <= width
        &&  item.y >= 0;
  });
}

function contains(a, obj){
  return a.findIndex( function(item) {
      return (item.x === obj.x && item.y === obj.y);
    }) !== -1;
}

function allowed_positions(width, pos, path) {
  var possible = possible_positions(width, pos);

  var allowed = [];
  var i = 0;
  for (i = 0; i < possible.length; i += 1) {
    if (!contains(path, possible[i])) {
      allowed.push(possible[i]);
    }
  }
  if (allowed.length === 0) {
      allowed = possible;
  }
  return allowed;
}

function next_pos(width, pos, path) {
  var allowed = allowed_positions(width, pos, path);
  var index = Math.floor(Math.random() * allowed.length);
  return allowed[index];
}
```

You will code the update of the pheromones shortly, but it doesn't affect your initial random paths. You've just built them. Let's see what they look like.

Showing the Trails

Draw the bag as a rectangle, using the canvas context:

```
var ctx = canvas.getContext("2d");
ctx.clearRect(0, 0, canvas.width, canvas.height);
ctx.fillStyle = "rgb(180, 120, 60)";
ctx.fillRect (edge, scale, canvas.width-2*edge, canvas.height-scale);
```

Don't make it the full canvas height—leave a scale gap for your ants at the top. You can draw each ant path as a line joining the spots visited. You need to scale up each coordinate and add the edge to the x value. Don't forget to take the y value from the height—the canvas has zero at the top, and you used zero for the bottom, as you did before on page 62. Use beginPath then moveTo the first position. Draw a lineTo to each point on the path, finishing with stroke to draw the line itself, as shown next:

Colonise/src/aco_paperbag.js
```
function draw_path(ctx, edge, height, path) {
  if (path.length === 0) {
    return;
  }

  var x = function(pos) {
    return edge + pos.x * scale;
  };
  var y = function(pos) {
    return height - pos.y * scale;
  };

  ctx.beginPath();
  ctx.moveTo(x(path[0]), y(path[0]));

  path.slice(1).forEach( function(item){
    ctx.lineTo(x(item), y(item));
  });
  ctx.stroke();
}
```

You can invoke setLineDash before each call to distinguish different paths, like this:

```
var was = ctx.setLineDash([5, 15]);
```

The numbers mean a line five units long, followed by a gap of fifteen, to give a dashed line.[1] An empty array sets it back to a solid line. You've got 25 ants, though, so you will run out of options. Instead, show the best and the worst paths once the ants are home to see if your ants learn better routes.

Iteratively Improve

Armed with a random setup, you can now help the ants iteratively improve. They learn in the aco function:

---

1. https://developer.mozilla.org/en-US/docs/Web/API/CanvasRenderingContext2D/setLineDash gives further details

```
Colonise/src/aco_paperbag.js
function aco(iteration, ants, pheromones, height, width) {
  var paths = new_paths(pheromones, height, width, ants);
  update(pheromones, paths, height);
  draw(iteration, paths);

  if (iteration === 50) {
    stop();
  }
  return iteration + 1;
}
```

As you can see, you need to make new_path for the ants, update the pheromones, and draw your results. This runs for 50 epochs. You can stop earlier if you don't see any improvements, or you can keep going until they find the best possible path. You get the idea.

Let's look at the pheromones first, since you need them to make a path. Store the pheromones in an array of objects with a position (x and y) and the weight for their actual level. To update the pheromones, you need to evaporate previous levels and add new values for each new path:

```
Colonise/src/aco_paperbag.js
function update(pheromones, paths, height) {
  evaporate(pheromones);
  paths.forEach( function(path){
    add_new_pheromones(height, pheromones, path);
  });
}
```

To evaporate you have several options; anything that makes the numbers shrink works. Dropping off by a fraction is straightforward:

```
Colonise/src/aco_paperbag.js
function evaporate(pheromones) {
  var rho = 0.25;
  for(var i = 0; i < pheromones.length; i += 1) {
    pheromones[i].weight *= (1-rho);
  }
}
```

Now add new pheromones for the latest paths. If your ants have never tried a point, you need to push a new pheromone, otherwise increase the current weight by L. To find L, use a constant Q to scale the reciprocal of path length. Bake double the height of the bag into L. You can experiment with other values too:

Colonise/src/aco_paperbag.js
```
function add_new_pheromones(height, pheromones, path) {
  var index;
  var Q = 2.0 * height;
  var L = Q/total_length(path);

  path.forEach ( function(pos) {
    index = pheromone_at(pheromones, pos);
    if ( index !== -1 ) {
      pheromones[index].weight += L;
    }
    else {
      pheromones.push( {x: pos.x, y: pos.y, weight: L} );
    }
  });
}
```

You need a helper function to find existing pheromones:

Colonise/src/aco_paperbag.js
```
function pheromone_at(pheromones, pos) {
  return pheromones.findIndex( function(item) {
    return (item.x === pos.x && item.y === pos.y);
  });
}
```

Make the total_length the sum of the Euclidean distances between each point on the path. You used this distance when you made a particle swarm on page 63:

Colonise/src/aco_paperbag.js
```
function euclidean_distance(first, second) {
  return Math.sqrt(Math.pow(first.x - second.x, 2)
                 + Math.pow(first.y - second.y, 2));
}

function total_length(path) {
  var i;
  var length = 0;
  for (i = 1; i < path.length; i += 1) {
    length += euclidean_distance(path[i-1], path[i]);
  }
  return length;
}
```

The ants use the pheromones to make new_path. For each ant, you create new_path by picking a starting spot and recording the next step until the ant goes above your bag, where you hid some food. Use the pheromone levels to decide where to move at each step probabilistically. When you made the initial paths, you randomly picked the next position; each possible place was equally likely. This time you want better places to be more likely:

Colonise/src/aco_paperbag.js

```
function pheromone_path(height, width, pheromones) {
  var path = [];
  var moves;
  var pos = start_pos(width);
  path.push(pos);

  while (pos.y < height) {
    moves = allowed_positions(width, pos, path);
    pos = roulette_wheel_choice(moves, pheromones);
    path.push(pos);
  }
  return path;
}
```

Call roulette_wheel_choice with the moves from the allowed_positions. Calculate the running total (i.e., partial_sum) of the functions of the pheromone, tau, and quality, eta, as taueta at these points, using the equation from *Using the Pheromones*, on page 82:

Colonise/src/aco_paperbag.js

```
function taueta(pheromone, y) {
  var alpha = 1.0;
  var beta = 3.0;
  return Math.pow(pheromone, alpha) * Math.pow(y, beta);
}

function partial_sum(moves, pheromones){
  var total = 0.0;
  var index;
  var i;
  var cumulative = [total];
  for (i = 0; i < moves.length; i += 1) {
    index = pheromone_at(pheromones, moves[i]);
    if (index !== -1) {
      total += taueta(pheromones[index].weight, pheromones[index].y);
    }
    cumulative.push(total);
  }
  return cumulative;
}
```

Make your selection by picking a random number between 0 and the overall total and return the corresponding spot:

```
     Colonise/src/aco_paperbag.js
Line 1  function roulette_wheel_choice(moves, pheromones) {
          var cumulative = partial_sum(moves, pheromones);
          var total = cumulative[cumulative.length-1];
          var p = Math.random() * total;
     5    var i;

          for (i = 0; i < cumulative.length - 1; i += 1) {
            if (p > cumulative[i] && p <= cumulative[i+1]) {
              return moves[i];
     10     }
          }

          p = Math.floor(Math.random() * moves.length);
          return moves[p];
     15 }
```

You find the totals on line 2 and pick a random number (p). Find the point this corresponds to on line 7. If your total is zero, just pick any possible move on line 13. You can now build up each ant path, and you should see the paths getting better over time.

Did It Work?

The ants move around the bag and find shorter paths. There are lots of parameters to play around with. The ants' starting position makes a big difference. Let's see what happens with 25 ants, running for 50 epochs, and dropping to pheromone level by 0.25 each time.

Starting in the Same Place

If the ants start in the same place, they will find the best path quite quickly. Some of the ants still wander around a little but do improve. A typical run gives about 30 steps for the worst initial path, 6 for the best, and an average of 15. The best path gets to the minimal 5 steps relatively quickly. You can see the ants find the best possible path in the figure on page 94. The best path is the solid line; the worst path is the dashed line:

Since your selection is probabilistic, you will see differences each time you run this. You can use the best, worst, and average path length to see what tends to happen. Add these elements to the HTML:

```
<p id="best">best distance</p>
<p id="worst">worst distance</p>
<p id="average">average distance</p>
```

and calculate your statistics like this:

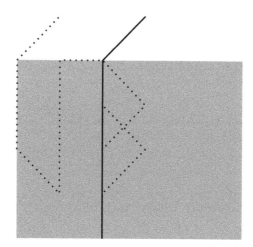

Colonise/src/aco_paperbag.js
```javascript
function find_best(paths) {
  var lengths = paths.map(function(item) {
    return total_length(item);
  });
  var minimum = Math.min(...lengths);
  return lengths.indexOf(minimum);
}

function find_worst(paths) {
  var lengths = paths.map(function(item) {
    return total_length(item);
  });
  var maximum = Math.max(...lengths);
  return lengths.indexOf(maximum);
}

function find_average(paths) {
  if (paths.length === 0) {
    return 0.0;
  }
  var sum = paths.reduce(function(sum, item) {
    return sum + total_length(item);
  }, 0);
  return sum / paths.length;
}
```

Then report your results, for example:

```javascript
document.getElementById("average").innerHTML = find_average(path);
```

If your ants are improving, you would expect your numbers to get smaller over time. Many machine learning algorithms need a statistical assessment to detect if they are learning. Let's see if things change when the ants start somewhere random.

Start Somewhere Random

If you use the same parameters as before, but let your ants start somewhere random they now have many more possible paths. They will still tend to find a relatively good path. However, the best ants often go diagonally rather than straight up, giving the path a kink, as the next picture shows:

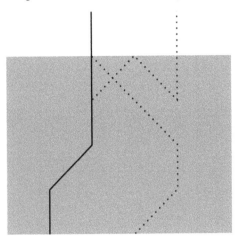

The worst and average lengths tend to be longer when you let the ants start anywhere. This should be no surprise—there are many longer paths to try. Over time, you will still see improvement. They can sometimes find a path straight up.

Alpha and Beta

Look back at the worst ant's path on page 94. Sometimes it heads the wrong way! Ants are attracted to spots on the best path, but your pheromones didn't encode any information about direction. Putting pheromones at the spots or nodes rather than edges allows—and sometimes encourages—ants to walk the wrong way down a good path. An edge has a start and end point, giving a direction. You could save the pheromones in a two-dimensional array, representing a matrix M, with M[i][j] storing the pheromone along the edge from i to j.

You did use the height (y) in your taueta function to give your ants a sense of direction. However, the pheromones count too. Remember:

```
function taueta(pheromone, y) {
  var alpha = 1.0;
  var beta = 3.0;
  return Math.pow(pheromone, alpha) * Math.pow(y, beta);
}
```

If you make beta zero in this function, Math.pow(y, beta) will always be 1, so only the pheromone contributes. Ants still find shorter paths, but overall the average distances tend to stay about the same. You can see the much longer worst path in the next picture:

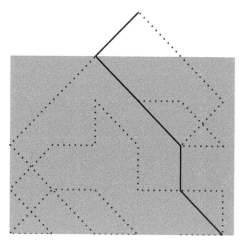

If you make alpha zero, the pheromones and therefore path lengths no longer influence the new ant path. The ants do still go up, with a few kinks, but the worst ant still wanders around all over the place as you can see in the next picture:

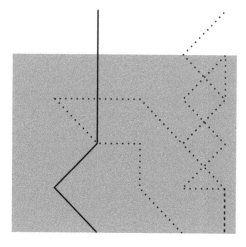

Using non-zero values gives a combination of your heuristic and path length. This gives you a colony; a zero only uses one aspect, so the ants go off doing their own thing.

Other Options

There are several parameters to experiment with in this algorithm. If a machine learning algorithm confronts you with lots of parameters, try making them zero to see what happens. Change them one at a time, and see what influence they have. Try different ρ values to control the evaporation, or change Q, which controls the influence of the path length. You can experiment with the number of ants, and you can change the scale or step size. If you use smaller numbers, the ants need to make longer paths, so may need to explore for longer. The next figure shows a simulation for smaller step sizes, using a scale of 10:

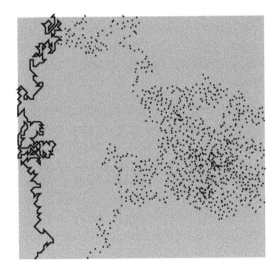

They have so many more possible paths now. They don't always find the best path, but they do improve over time.

Over to You

You made an ant colony optimization and it does sometimes find the best possible path. Of course, you knew in advance the best path out of the bag. You had several parameters to fiddle with. You can try to optimize these if you track what happens as you change them. You could even use a genetic algorithm to try to find the best combination.

There are many ways to extend or change your algorithm:

- Change when the updates happen to see if the ants learn more quickly

- Use daemon actions to lay extra pheromones along better paths once in a while

- Try the TSP by letting your ants visit any other spot at each move, but only visiting each spot once per journey

- Kill off any ants that get stuck in a corner instead of letting them revisit a previous point

- Get your computer to play the old arcade game Snake! To make a longer path, use the path length in taueta instead of its reciprocal. Watch out for the snake hitting itself.

Never be afraid to ask, "Why?" or try variations to see what happens.

You've seen what's officially known as a simple ACO in this chapter. If you want to read more about ACO, have a look at Stutzle and Hoos' newer variant called the Max-Min ant system.[2] This is likely to perform better for more complicated problems.

You have seen two nature-inspired swarm algorithms, moving agents around space. The PSO concentrated on moving the agents out of the bag by encouraging them to go up. In contrast, the ACO focused on building a shorter path. In the next chapter, you will move particles about, driven by a model. Each step will be random, though the model will ensure your particles diffuse out of the paper bag over time. This will give you a taste of *Monte Carlo* simulations. For some simulations, you always get the same outcome. Other times, like tossing a coin, will give you a different sequence of events. You expect as many heads as tails on average, but can't be sure exactly what order they will happen in. You will see more details in the next chapter. You will also see how property-based testing works. This will add to your knowledge of machine learning techniques and give you more ways to assess if your algorithms work.

CHAPTER 6

Diffuse! Employ a Stochastic Model

In the previous chapter, you built an ant colony optimization. The ants found trails out of your paper bag by sharing information about the lengths of the routes they took. You may not care which path an ant takes if you only want to code your way out of a paper bag. When ants, particles, or points spread out, they end up outside the paper bag. Problem solved. A *simulation* shows what happens when things spread out using a model or equation of how they spread or *diffuse*.

A simulation lets you explore different scenarios. Simulations are used in a variety of areas from finance to epidemiology. Armed with a plausible model —often a *stochastic differential equation* (SDE)—a simulation shows three things: worst case scenarios, how likely something is to happen, and what the outcome might be if you change the parameters:

- What if using mosquito nets decreases the chance of malaria spreading by 5%?

- What if the interest rate increases by 0.25%?

- What if the interest rate drops below zero?

This chapter builds three stochastic models of diffusion in C++ using the random numbers in the standard library. This gives you a very different flavor of machine learning, adding to your repertoire. The simulations model *Brownian motion*, starting with a type of random walk known as Markov processes. Many machine learning algorithms use Markov processes, so it's worth being familiar with the term.

Imagine releasing a cloud of particles into the middle of a paper bag. Over time, they spread out and diffuse, eventually going through the edges of the bag. The formula for Brownian motion is a good model for this. It has a random element, so you'll make a *Monte Carlo simulation* showing how particles might travel. Every simulation run will be slightly different, but the particles still diffuse. You can change the model a little to explore how stock prices might change over time, and see what happens as the interest rate changes.

By the end of this chapter, you will be comfortable creating simulations and confident with terms like stochastic and Monte Carlo simulation. You'll use a media library to draw particles diffusing, and learn about property-based testing. Any code with a random element can be hard to test. Property-based testing checks overall properties, without worrying about checking for specific runs of numbers.

Your Mission: Make Small Random Steps

Let's define Monte Carlo simulations first, then see how Brownian motion will diffuse particles out of a paper bag. Geometric Brownian motion will build on the first model, giving a simulation of stock prices. Then you can sneak in potential price jumps, giving three models to simulate.

These models are stochastic. When you built a genetic algorithm in Chapter 3, *Boom! Create a Genetic Algorithm*, on page 33 you used a deterministic model—you determined the exact path of the cannonball for a given angle and velocity. In contrast, a stochastic model has a random element. You don't know in advance exactly what will happen; however, you can work out properties, including an average or expected result, and how much variation occurs over several runs.

Monte Carlo Simulations

A Monte Carlo simulation investigates numerical problems that cannot be solved directly, giving answers with varying degrees of accuracy. The name deliberately invokes images of casinos and gambling. Let's consider an example by trying to find the area under a curve.

If the curve has an equation which is integrable, you can use *calculus* to find the area under the curve. However, if the curve is a hand-drawn squiggle, you might have trouble finding the function describing the curve, let alone doing the math. To help, you can use an estimation scheme.

Consider a hand-drawn curve, something like the one in the picture. Try to find the area in the curve:

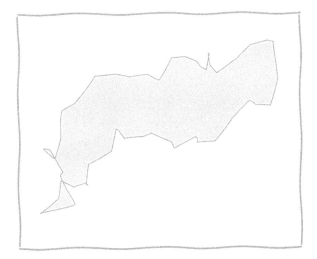

If a grid is superimposed, the area can be estimated by counting how many unit squares contain a portion of the curve. Notice that fewer than 19 whole squares in the next picture contain the curve, giving an upper bound for the area. Finer grained grids will give more accurate estimates:

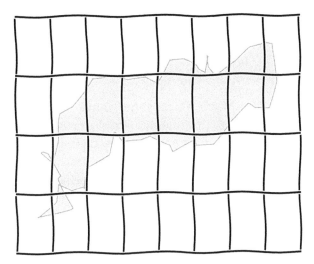

Alternatively, you can throw darts at the paper and count how many are inside the curve. There are 30 darts in the next picture:

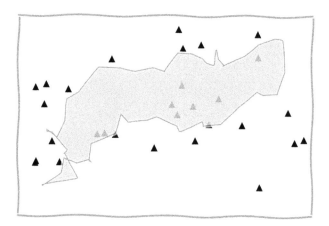

Ten darts hit some portion of the curve. This works out to be 10/30, or 1/3 inside the curve, giving approximately 33% of the area of the rectangle. Several such experiments will inevitably give varying areas, which can be averaged or used to give a lower and upper bound. The essence of any such simulation is the same: run an experiment a few times and see what happens. Try out this area finding method. If you choose a rectangle, perhaps a paper bag, or another shape you can calculate the area of easily, write code to pick several random points and see how many are inside the rectangle or not. Then take the average. You can work this through without help.

Let's see how to make a Monte Carlo simulation of something a bit more complicated: diffusion, using Brownian motion, Geometric Brownian motion, and Jump Diffusion. Rather than a static rectangle, you will see elements moving over time, driven by a model or equation. How do they move? Each different model has the same essence, moving by taking a step from the current state or position. These are described by equations of the same form:

```
next_position = current_position + f(parameters)
```

The function f varies for each model. In each case, you get a sequence of positions to show particles or stock prices moving over time.

Brownian Motion

A diffusing substance moves, apparently randomly, from places of higher concentration to those with lower concentration, eventually reaching equilibrium. There are various diffusion equations. They can be at the molecular level, in solids, liquids, and gases, driven by pressure, temperature, or electrical energy. The diffusion can involve turbulence, stirring the liquid, or spinning up a turbine. The simplest model is *Brownian motion*. This models small particles bouncing off tiny liquid or gas molecules. They move around independently, and you'll ignore the particles bouncing off one another. By modeling several particles moving, they spread out or diffuse over time.

Brownian motion models particles moving by small steps. To get an even spread, each direction needs to be equally likely. If we make the mean step zero, there will be no drift, but the particles will still spread out. To ensure a substance diffuses, the variance of the steps needs to be big enough, but not too big. Let's think about the steps, the mean and the variance to find a suitable equation for this model.

 Joe asks:

What's Mean and Variance?

The arithmetic mean is one type of average. Find the total of the values and divide by how many values you have. For n values this is:

$$\text{mean} = \frac{\sum_{i=1}^{n} \text{value}_i}{n}$$

The variance measures how far from this mean your values are. With a mix of positive and negative values, some will cancel out when you add them up. If you square the numbers, you get positive values so none get cancelled;

$$\text{variance} = \frac{\sum_{i=1}^{n} (\text{value}_i - \text{mean})^2}{n}$$

Take the square root to find the standard deviation.

Each particle will move a small step in any direction. This creates a special type of *random walk*. Some random walks use the last few moves to drive the next move, perhaps avoiding a previously visited spot. With this simulation,

Joe asks:
What's a Normal Distribution?

Plotting people's heights in groups of 10 cm gives a curve where few people are very short or very tall. Most are somewhere in the middle, giving a bell-shaped curve, shown in the figure. As you shrink the range down from 10cm, the histogram tends toward a symmetric shape shown in the figure. This can be modeled by the Gaussian function:

$$f(x) = \frac{1}{\sqrt{2\pi\sigma^2}} \times \exp-\frac{(x-\mu)^2}{2\sigma^2}$$

where σ^2 (sigma sqaured) is the variance and μ (mu) is the mean. Strictly speaking, the area of the bars in the chart tend to the area under the curve. This is also called a Gaussian distribution.

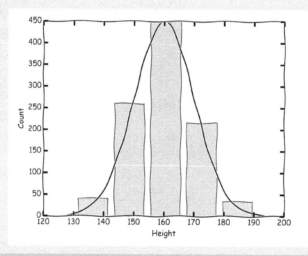

a particle move only depends only on where it is now, rather than the last few positions, so is memoryless. This makes your random walk a *Markov* chain or process. Markov chains concern sequences of events. They are like state machines, but the next step is picked at random. When you search for something, does your search engine suggest the next words you might type? Does predictive text guess your next words? These are sequences of events, so can be built with Markov chains. Sometimes the states are not visible, giving you a *hidden Markov model*.[1] These crop up in many machine learning contexts, so it's worth knowing the term.

---

1. en.wikipedia.org/wiki/Hidden_Markov_model

As well as being a Markov process, simple Brownian motion has a *mean* step size of zero. This makes left as likely as right—up as likely as down. You'll try adding a drift later, making one direction more likely. The *variance* of the steps is a multiple of the time step. This ensures spread or diffusion occurs —too small, and particles clump together; too big, and they zoom off. You get the required properties if you make the steps *normally* distributed. Picking them from the std::normal_distribution will give you exactly what you need.

To get an intuitive sense of this, consider a simple random walk along a line. Start at an origin (0) and imagine tossing a coin. Heads means go left (-1), tails means go right (+1).

Find the mean and variance for several walks. The random walk equation is:

```
next_position = current_position + pick_one_of(-1, 1)
```

For several single step walks, you get an average of 0 steps. You expect to go right (+1) as often as you go left (-1) in the long run. For walks with more steps, you still get an average of 0, because you can add the average of the single step walks.

For single step walks, you get a variance of 1 whether you go right or left. Why? Any single step walk has a mean of 0, so the variance of n walks is

$$\frac{\sum_{i=1}^{n}\left(\text{value}_i - 0\right)^2}{n} = \frac{\sum_{i=1}^{n}\left(\pm 1 - 0\right)^2}{n}$$

Since

$$\left(\pm 1 - 0\right)^2 = 1$$

the variance must be

$$\frac{\sum_{i=1}^{n} 1}{n} = \frac{n}{n} = 1$$

For longer walks you can add the variance of each step. For example, for walks of four steps you get a total variance of four:

$$1 + 1 + 1 + 1 = 4$$

The variance is the same as time step. This works because the steps are independent.[2]

---

2. stats.stackexchange.com/questions/159650/why-does-the-variance-of-the-random-walk-increase

To see these means and variances, you need to run enough simulations. A few will give you a feel for what happens. For a real-world simulation, you may need to run several thousand or even million to be *confident* in the numbers. The precise number follows from your setup.[3]

By using random numbers with a normal or Gaussian distribution, you simulate Brownian motion. You then get a varying step size, rather than taking steps of exactly one unit each time. Armed with a particle at some point (x, y) and a source of independent random numbers called ΔZ_1 and ΔZ_2 move the particle to

$$(x + \sqrt{\Delta t}\, \sigma \Delta Z_1,\ y + \sqrt{\Delta t}\, \sigma \Delta Z_2)$$

in each time step (Δt) . These differences

$$\Delta x = \sqrt{\Delta t}\, \sigma \Delta Z_1$$

$$\Delta y = \sqrt{\Delta t}\, \sigma \Delta Z_2$$

are *stochastic differential equations* (SDE). The Greek letter Delta signifies a difference. Stochastic means something random is hiding in there. The Zs are Gaussian random numbers, with a mean of 0 and variance of 1. Sigma is a variance of your choice, scaling up the steps. This gives the step to add to the current position:

```
next_x_position = current_x_position + dx
next_y_position = current_y_position + dy
```

To implement this, make two independent draws from a normal random number generator, one for the dx step, and one for the dy step.

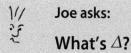

Joe asks:

What's Δ?

Calculus uses various letters; $\delta,\ \Delta,\ d,\ \partial$ to represent a difference, change, or rate. Δ gives a discrete step or difference, while d gives the instantaneous difference. If you're not familiar with calculus, treat Δ as a step or change in each iteration. The d is the limit as the step size gets smaller and smaller.

---

3. https://stats.stackexchange.com/questions/34706/simulation-study-how-to-choose-the-number-of-iterations

Geometric Brownian Motion

You can build on this first random walk using a different diffusion model to simulate stock prices. Armed with a starting price and model, a simulation gives prices over time. You get a spread of possible prices which you can plot inside a paper bag, and some might go above the paper bag. The y-coordinate will be a fictitious stock price (S) and time (t) will be the x-coordinate. You can see the possible price curves spreading or diffusing if you plot them side by side. To model this you will use *Geometric Brownian motion* (GBM). This is very similar to the first model; however, the logarithm of the steps follow Brownian motion rather than the steps themselves.

Geometric Brownian motion uses a different equation but still models steps being taken over time.

```
next_price = current_price + price_change
```

Last time, you found `dx` to add to `x`, and you found `dy` to add to `y`, making particles move in space. Now you choose a time step and find the corresponding price change from a model. Let's use this SDE to model stock prices moves:

$$\Delta S = S \times (\mu \Delta t + \sigma \Delta W)$$

This tells you the price difference ΔS to add to the current price S. There are other models, but this is relatively common. Like the ΔZ_1 and ΔZ_2 before, ΔW is a draw a number from a Gaussian random number generator. We will call this `dW` in code.

The drift μ in this equation models the return on your investment. You also have a scale parameter σ, usually called *volatility*. It relates to the variance of the step sizes—larger values allow larger stock price movements at any given moment. If this is zero, the stochastic part of the model stops, modeling the returns from a completely secure investment. Any simulation will give the same prices. If this is non-zero, your investment may go up or down at any moment, but it will drift up on average by μ. Some simulations will give higher or lower prices, so you see a spread between price curves increasing over time. You use these parameters to set up your `std::normal_distribution`. By default, it uses 0 for the mean and 1 for the standard deviation.

To build your stock price simulation, you need an initial stock price, a drift, volatility, and a source of random Gaussian numbers `dW`. You will then generate a sequence of possible stock prices after each time step `dt`. You will plot these as a stock price curve, using your paper bag as axes. Instead of particle doing random walks, you now have points on a curve. Time gives you the x value, and the price a y value. The bottom of the bag is 0, and you will start the stock price somewhere above zero.

Why above zero? Why not at zero? If the stock is initially zero, each price step will be zero since

$$\Delta S = 0 \times (\mu \Delta t + \sigma \Delta W) = 0$$

Any initial stock value greater than zero will do. The bottom of the bag represents the time your model runs, starting with zero on the left. Let's imagine the simulation occurs over a two week time period (or whatever time period you like). You will need to choose the time steps (dt) between simulated prices, as well as its drift and volatility. More steps give you more points. If you choose the right parameters, your line of stock prices ends up above the bag.

Jump Diffusion

Both Brownian motion and Geometric Brownian motion are *continuous* models—the particles, or stock prices, do not teleport to somewhere completely different. Furthermore, if you zoom in or take stock prices at shorter intervals, the overall shape or diffusion will look very similar. This fits some situations well, but sometimes a process can change or jump. The stock market might crash, or instead of tanking, start to soar.

A key property of the Brownian motion model is its *continuity*. This is a precise mathematical concept, but when a line can be drawn without taking a pen off the paper, it is considered *continuous*. In contrast, a *discontinuous* path will have a jump—a point where the path breaks and picks up elsewhere— making it look rather more like two lines! Introducing jumps into the simulation causes *discontinuities*. Your first model only allows small stock price changes. Your jump model allows occasional large jumps. Sometimes the jumps will drive the numbers up; sometimes they might drive the numbers down. If you cheat and force these jumps to be positive, you are more likely to escape the bag.

You can use the *Poisson* distribution to simulate something happening occasionally. Draw a number ΔN from the Poisson distribution to decide how many jumps happen in a time step Δt and decide the jump size J. Adding this extra term to the last model gives Jump Diffusion:

$$\Delta S = S(\mu \Delta t + \sigma \Delta W + J \Delta N)$$

This price change, ΔS tells you the step to the next price as before:

```
next_price = current_price + price_change
```

When ΔN is 0, this collapses to the previous model; when it is non-zero you have a discontinuity or jump. You can code these together, making the jump size zero if you want plain Geometric Brownian motion without jumps. Make it non-zero for jumps.

\\//
ᔑᓵ **Joe asks:**
What's a Poisson Distribution?

Think about the time spent waiting for a bus. Sometimes the wait is very short; usually, it's a little while; and now and then, it seems to take forever. The Poisson distribution models how many times an event happens in a time period. The shape of this distribution comes from the function

$$f(x) = \frac{\lambda^n e^{-\lambda}}{n!}$$

where λ is the rate of the event for which you are waiting, and n is how many times it happens. If you count how many events happen in a time interval, say how many buses turn up, to make a bar chart, its area tends to the area under this curve, as suggested in the figure.

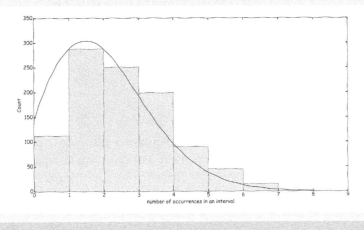

How to Cause Diffusion

You now have an overview of how to build Brownian motion, Geometric Brownian motion, and Jump Diffusion. This section will show you how to get random numbers for the stochastic part of the simulation, and how to draw pictures in C++.

Small Random Steps, dW

C++ introduced a random number library from C++11, with a variety of statistical distributions including the Gaussian and Poisson flavors you need. You will therefore have this avaiable, without needing to install a library, if you use a current compiler—for example, anything from GCC4.8.1.[4] You can use the older rand C call instead, but this is error prone and you will likely need to resort to some tricks to get the distributions you need. The new standard C++ library is much easier to use—simply include the header.

Imagine simulating dice rolls. You need an integer between 1 and 6, each having the same chance of being rolled. In code, include the random header and make an engine to drive your distribution, giving it a seed. Using the same seed produces the same run of numbers each time while using a different value each time produces different numbers. The standard random header provides std::random_device you can use as a seed. It is supposed to produce non-deterministic random numbers. Be forewarned, it may not work on your setup. Test it![5]

Include the random header and simulate rolling a die like this:

```
int main()
{
  std::random_device rd; //or seed of your choice
  std::mt19937 engine(rd());
  std::uniform_int_distribution<> distribution(1, 6);
  int die_roll = distribution(engine);
}
```

You call the distribution using the engine to get a any number from 1 to 6. You have simulated a die roll. To make your simulations, you need a std::normal_distribution for the dW step and std::poisson_distribution for the jumps. You will also need a way to display the results.

Drawing in C++

There are many options for drawing in C++. This chapter will use the Simple and Fast Media Library (SFML).[6] You will need the library built for your partiucular operating system and toolchain. You will also need to add the library and path to the headers in your project or makefile. The tutorials on the library website are there if you need help. Don't worry! You can still run the

---

4.   isocpp.org/wiki/faq/cpp11

5.   en.cppreference.com/w/cpp/numeric/random/random_device

6.   www.sfml-dev.org/

simulations if you have a different library, or you can simply stream out the numbers—though that won't be as much fun to watch!

Once installed, create a main file and include the SFML/Graphics.hpp header. Create make a window on which to draw, with a size and a title. Then loop while this is open, checking for events like the window closing. If it's still open, clear the window, redraw what you need to and then call display. That's it!

```
int main()
{
  sf::RenderWindow window(sf::VideoMode(200, 200), "Hello, world!");

  while (window.isOpen())
  {
    //check for events here, like window closed

    window.clear();

    //draw again here

    window.display();
  }
}
```

You will draw a bag using a sf::RectangleShape for the edges. You can use a sf::CircleShape for each particle in your first simulation. These particles diffuse in each direction, and some get out of the bag over time. Next, find some stock prices over time and join the dots using sf::Vertex to draw a line between the points. These prices start on the left and move up, randomly. If you plot several simulations side by side you will see a spread or "diffusion" over time. Some stock prices go above the bag. For each simulation, you update the visualization in the while loop, so you can see the movement.

Let's Diffuse Some Particles

You have now seen how to make a Monte Carlo simulation of three different stochastic differential equations. The first is Brownian motion. The second and third are Geometric Brownian motion. These will simulate stock prices, first without jumps, then with jumps. You can use the same code for the stock prices, making the jump size zero if you don't want any jumps. Let's code it.

Brownian Motion

You need particles to move, so code a Particle class with a position (x, y) and a way to Move. To avoid busting through the sides of the bag, specify the bag's edges and its size. When a particle is high enough to escape from the bag, it's done and stops moving:

```cpp
class Particle
{
public:
  Particle(float x = 0, float y = 0, float edge = 0,
      float max_x = std::numeric_limits<float>::max(),
      float max_y = std::numeric_limits<float>::max(),
      bool breakout = false)
    :
    x(x), y(y), edge(edge),
    max_x(max_x), max_y(max_y),
    done(false),
    breakout(breakout)
    {
    }

  void Move(float x_step, float y_step)
  {
    if (done) return;

    x += x_step;
    y += y_step;

    if (y < edge / 4)
    {
      done = true;
      return;
    }
    if (y > max_y) y = max_y;

    if (!breakout)
    {
      if (x < edge / 2) x = edge / 2;
      if (x > max_x) x = max_x;
    }
  }

  float X() const { return x; }
  float Y() const { return y; }
private:
  float x;
  float y;
  const float edge;
  const float max_x;
  const float max_y;
  bool done;
  const bool breakout;
};
```

The particles move as they bump into the air. Use the std::normal_distribution to generate the bumps. Use floats for the SFML and scale up by your step to get the Bump:

Diffuse/Lib/Air.h
```
class Air
{
  std::mt19937 engine;
  std::normal_distribution<float> normal_dist;
  const float step;

public:
  Air(float step,
    unsigned int seed = 1)
    :
    step(step),
    engine(seed)
  {
  }

  float Bump()
  {
    return step * normal_dist(engine);
  }
};
```

That's it! Let's run a simulation.

Make a std::vector to store some particles. Don't forget to include its header, along with your Air and Particle. Decide how many particles you want and where they start (start_x and start_y). Also choose the bag height and width, distance to the edge of the window, and whether or not particles should breakout of the sides. Choose a lineWidth for the thickness of your bag so you can avoid particles nudging into the sides. Add them to your vector in a loop:

Diffuse/MC101/particle_main.cpp
```
std::vector<Diffuse::Particle> createParticles(size_t count,
  float start_x,
  float start_y,
  float lineWidth,
  float edge,
  float height,
  float width,
  bool breakout)
{
  std::vector<Diffuse::Particle> particles;
  for (size_t i = 0; i < count; ++i)
  {
    particles.emplace_back(
      start_x,
      start_y,
      edge + lineWidth,
```

```
        edge / 2 + width - 2 * lineWidth,
        edge / 2 + height - 2 * lineWidth,
        breakout
        );
    }
    return particles;
}
```

Now that you have some Particles, you need some Air. Decide the step size and seed. You can use random_device if it works on your setup. Decide how many simulations you want and loop around; Bump the x and y coordinates each time:

```
const float step = 7.5f;
std::random_device rd;
Diffuse::Air air(step, rd());

for (int i=0; i<sims; ++i)
{
    particle.Move(air.Bump(), air.Bump());
}
```

This doesn't give you much to see! Use the SFML introduced in the drawing code on page 111 to draw your particles. Add the libraries required to your build and write an action function. Make a window for drawing, create some air and some particles, say 25, and make them Move in a loop:

```
Diffuse/MC101/particle_main.cpp
Line 1  void action(size_t count, float step, bool breakout)
        {
            std::stringstream title;
            title << "2D Brownian motion " << count << ", breakout " << breakout;
    5
            const float height = 500.0f;
            const float width = 500.0f;
            const float edge = 50.0f;
            const float lineWidth = 5.0f;
    10      const auto bagColor = sf::Color(180, 120, 60);

            int max_x = static_cast<int>(width + edge);
            int max_y = static_cast<int>(height + edge);
            sf::RenderWindow window(sf::VideoMode(max_x, max_y),
    15                  title.str());

            std::vector<Diffuse::Particle> particles =
                createParticles(count, max_x/2.0f, max_y/2.0f,
                            lineWidth, edge,
    20                      height, width, breakout);

            std::random_device rd;
            Diffuse::Air air(step, rd());
```

```
25    bool paused = false;
      while (window.isOpen())
      {
        sf::Event event;
        while (window.pollEvent(event))
30      {
          if (event.type == sf::Event::Closed)
            window.close();
          if (event.type == sf::Event::KeyPressed)
            paused = !paused;
35      }

        window.clear();

        drawBag(window, lineWidth, edge/2, height, width, bagColor);
40
        sf::CircleShape shape(lineWidth);
        shape.setFillColor(sf::Color::Green);
        for(auto & particle: particles)
        {
45        if (!paused)
            particle.Move(air.Bump(), air.Bump());
          shape.setPosition(particle.X(), particle.Y());
          window.draw(shape);
        }
50      window.display();
        std::this_thread::sleep_for(std::chrono::milliseconds(100));
      }
    }
```

You can pause or restart the diffusion with a key in the event loop as shown on line 28. When the simulation is running, the air moves the particles around. Draw the bag using rectangles for the left, bottom, and right edges. You don't need this for the simulation, but it shows the particles spreading or diffusing clearly:

```
Diffuse/MC101/particle_main.cpp
void drawBag(sf::RenderWindow  & window,
  float lineWidth,
  float edge,
  float height,
  float width,
  sf::Color bagColor)
{
  sf::RectangleShape  left(sf::Vector2f(lineWidth, height));
  left.setFillColor(bagColor);
  left.setPosition(edge, edge);

  sf::RectangleShape  right(sf::Vector2f(lineWidth, height));
  right.setFillColor(bagColor);
  right.setPosition(edge + width, edge);

  sf::RectangleShape  base(sf::Vector2f(width + lineWidth, lineWidth));
```

```
    base.setFillColor(bagColor);
    base.setPosition(edge, edge + height);

  window.draw(left);
  window.draw(right);
  window.draw(base);
}
```

You have built a Monte Carlo simulation. You can make this into a stock price simulation by changing the step size. These two very different sounding simulations have similar building blocks. The first moves particles in space. The second moves stock prices over time. They both add a random step each time. The particles spread out over time, all centered around the starting point due to the zero mean of the random walk. The stock prices will spread out or diverge over time as well. If you try several simulations, some end higher than others, but on average they will go up by the same drift.

Stock Prices

Code stock prices with and without jumps together. Use a non-zero jump if you want potential jumps; use zero to turn them off. Make a PriceSimulation class to generate the Next stock prices:

Diffuse/Lib/PriceSimulation.h
```
class PriceSimulation
{
public:
  PriceSimulation(double price,
    double drift,
    double vol,
    double dt,
    unsigned int seed =
      std::chrono::high_resolution_clock::now().
                    time_since_epoch().count(),
    double jump = 0.0,
    double mean_jump_per_unit_time = 0.1);

  double Next();

private:
  double price;
  double drift;
  double vol;
  double dt;
  double jump;

  std::mt19937 engine;
  std::normal_distribution<> normal_dist;
  std::poisson_distribution<> poisson_dist;
};
```

Set up an engine for the random numbers in your constructor, using the default normal distribution, with a mean of 0 and standard deviation of 1. Give your Poisson distribution a rate at which jumps happen and scale it by the time step (dt). Make jump zero to stop the jump, or a positive number to make the prices jump up:

Diffuse/Lib/PriceSimulation.cpp

```
PriceSimulation::PriceSimulation(double price,
    double drift,
    double vol,
    double dt,
    unsigned int seed,
    double jump,
    double mean_jump_per_unit_time)
  :
  price(price),
  drift(drift),
  vol(vol),
  dt(dt),
  engine(seed),
  jump(jump),
  poisson_dist(mean_jump_per_unit_time * dt)
{
}
```

Simulate the Next price using a stochastic step built from a drift over time, a jiggle, and possibly a jump as follows:

Diffuse/Lib/PriceSimulation.cpp

```
Line 1  double PriceSimulation::Next()
     2  {
     3      double dW = normal_dist(engine);
     4      double dn = poisson_dist(engine);
     5      double increment  = drift * dt
     6        + vol * sqrt(dt) * dW
     7        + jump * dn;
     8      price += price * increment;
     9      return price;
    10  }
```

The jiggle dW comes from your normal distribution on line 3, and the jumps come from the Poisson distribution on line 4. Sum these to get the price movement, shown on line 7. Multiply to get your stock price change, shown on line 8. Call this in a loop to get a simulation of possible future stock prices from your chosen starting price.

Choose the drift, volatility, and jump size. The time is the bag width. Use the time step dt to choose the number of prices; smaller numbers give you more. Run your simulation and record the prices you get back so you can plot these:

Diffuse/StockPrice/stock_main.cpp

```cpp
std::vector<sf::Vertex> price_demo(unsigned int seed,
  double drift,
  double vol,
  double time,
  double dt,
  double jump,
  double mean_jump_per_unit_time)
{
  const double start_price = 50.0;
  Diffuse::PriceSimulation price(start_price,
    drift,
    vol,
    dt,
    seed,
    jump,
    mean_jump_per_unit_time);

  std::vector<sf::Vertex> points;
  const int count = static_cast<int>(time/dt);
  points.push_back(sf::Vector2f(0.0f, static_cast<float>(start_price)));
  for(int i=1; i <= count+1; ++i)
  {
    auto point = sf::Vector2f(static_cast<float>(i*dt),
      static_cast<float>(price.Next()));
    points.push_back(point);
  }
  return points;
}
```

To plot these using SFML, make a sf::Vertex for each stock price, and join them with a line. Draw the bag as you did before. Your stock price starts at time zero and updates at each time step(dt). You want the stock simulation to run over the width of the bag so scale the time values, so they reach the other side of the bag. To get the height of each point, subtract the stock price from the height of your window. Remember 0 is at the top, but you want it to be at the bottom. Pull it all together, like this:

Diffuse/StockPrice/stock_main.cpp

```cpp
void action(const std::vector<std::vector<sf::Vertex>> & sims,
  float time,
  float height,
  std::string title)
{
  const float edge = 30.0f;
  const float lineWidth = 5.0f;
  const float width = 500.0f;
  const float x_scale = width/time;
  const auto bagColor = sf::Color(180, 120, 60);
  sf::RenderWindow window(
```

```
    sf::VideoMode(static_cast<int>(width + 2*edge),
    static_cast<int>(height + 2*edge)),
    title);

size_t last = 1;
while (window.isOpen())
{
  sf::Event event;
  while (window.pollEvent(event))
  {
    if (event.type == sf::Event::Closed)
      window.close();
      break;
  }

  window.clear();

  drawBag(window, lineWidth, edge, height, width, bagColor);

  last = std::min(++last, sims.begin()->size() - 1);
  for(const auto & points: sims)
  {
    bool out = false;
    for(size_t i=0; i < last; ++i)
    {
      auto scaled_start = sf::Vertex(
        sf::Vector2f(points[i].position.x * x_scale + edge,
        height - points[i].position.y),
        sf::Color::White);
      auto scaled_point = sf::Vertex(
        sf::Vector2f(points[i+1].position.x * x_scale + edge,
        height - points[i+1].position.y),
        sf::Color::White);
      sf::Vertex line[] = {scaled_start, scaled_point};
      window.draw(line, 2, sf::Lines);
    }
  }
  window.display();
  std::this_thread::sleep_for(std::chrono::milliseconds(50));
  }
}
```

Try it out with different parameters. Try it with the jumps on and off. Vary the jump size. What drift do you need to escape the paper bag? Make the volatility zero and notice the exponential curve you get. Turn up the volatility and see what happens. Most importantly, have fun with it!

Did It Work?

You now have three Monte Carlo simulations of Brownian motion. The first gradually moves some particles out of the bag. They start in the middle and gradually disperse as the figures show. They will either bump into the sides or breakthrough depending on the value you used for breakout. If they are allowed through the sides, some will sneak back in again from time to time. Because they stop moving when they get above the bag, you will end up with a line near the top of the window, as shown in the figure on the right:

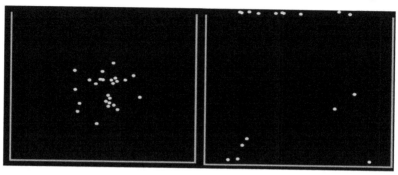

The stock price simulation has many more parameters. A simulation with time steps of 0.01, a drift of 0.2 (yes, a 20% return on your investment!), and zero volatility gives you prices rising over time but never getting out of the bag. The figure on the left shows this slight price rise. Make the drift 50%, keeping the volatility at zero, to go over the edge of a square bag as the image on the right shows:

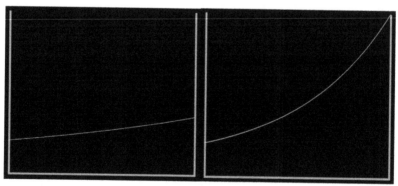

Without volatility, there is no random movement. Turn on the volatility and see what happens. Try a few simulations at once. You should get some shapes that look reminiscent of stock prices moving over time, similar to the figure on the left coming up. Add in some jumps—if they are positive, the price will only jump up, so it's more likely to get out of the bag. The next image shows five simulations, with a drift of 50%, and 10% (0.1) volatility—first without jumps, then with jumps of 0.5, with a probability of 0.25. Because this is random, the precise values will differ per simulation, though all the curves should drift up. The curves go straight up when the price jumps, seen in the figure on the right:

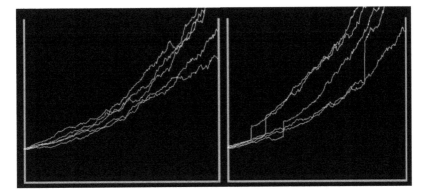

The simulations look believable, but are you sure they are right? Eyeballing your plots might help you find catastrophic errors, such as no points at all. You have points, and they look about right. This is not compelling enough though. Even with unit tests for the code, are you sure your code does the right thing?

Property-Based Testing

Let's consider another way to verify your code. Whenever a solution uses random numbers, it is challenging to test. Using a known seed, rather than random_device or similar gives a fixed sequence of numbers. This gives one way to check for regression bugs. If the output changes, something broke. However, how do you check your code does the right thing in the first place?

With the randomness, the exact values for each simulation differ, though certain properties will be the same. Can you think of any properties for these simulations? What happens if a stock price starts at zero? You can use properties like this to test your code. Let's see how, starting with a unit test and building this into a *property-based test*.

For these models, you chose a mean and variance and other parameters. You cannot write unit tests for every possible floating-point number! Unit tests can miss problems. You could randomly try a few numbers, and report any you find that don't have your required properties. We'll use a property-based testing library to do this.[7] It's header only, so simple to use—you clone it from

---

7.   software.legiasoft.com/git/quickcheck.git

the Git repo and include the header. Many others exist and some are better than others.[8] What matters is the idea, not the tool you pick.

Property-based testing randomly picks inputs and reports back if any break a stated property. Some go beyond picking random inputs by using something like a fitness function to seek out bad values. You know about this now, so might even be able to write your own. The Haskell package QuickCheck is held up as the exemplar of property-based testing.[9] It was built in the late 1990s, and many languages now have their own variants, so you can probably find a library for your language of choice. Let's use a version of QuickCheck implemented in C++. We'll start with a unit test for a stock price starting at zero and see how to generalize this to a property test.

Without jumps, a stock price starting at zero stays at zero as you saw on page 108. Let's test this using *Catch*.[10] If you haven't used Catch before, mentally translate it into whichever testing framework you use.

To build tests, tell it to generate a main for you:

```
#define CATCH_CONFIG_MAIN
```

and include the catch.hpp header. The tests themselves take a name and a tag and have assert macros like REQUIRE. Here's a test for zero stock prices staying at zero:

Diffuse/UnitTests/UnitTests.cpp
```
TEST_CASE("A stock price simulation starting at 0 remains at 0", "[Property]")
{
  const double start_price = 0.0;
  const double dt     = 0.1;//or whatever
  const unsigned int seed = 1;//or whatever
  Diffuse::PriceSimulation price(start_price, 0.3, 0.2, dt, seed);

  REQUIRE(price.Next() == 0.0);
}
```

REQUIRE checks the Next price is still 0.0. The previous lines set up the simulation. In addition to the start_price you want to pin to zero, the simulation has a few other parameters. What do you use for these? Hardcoded magic numbers? How many times have you missed an edge case when you've done that?

Property-based testing frameworks have *generators* to pick numbers, or strings or any other types, for you. The C++ QuickCheck library has a generator for built-in types like float which you use to make a generator for your own types.

---

8. github.com/emil-e/rapidcheck
9. hackage.haskell.org/package/QuickCheck
10. github.com/philsquared/Catch.git

Filter out items generated with unacceptable parameters, for example, a non-positive time step, so they don't get used in the tests. Finally, spell out the Property you want to check. For a stock price starting at zero, you want all the generated prices to be zero. Let's work through this.

Include the header quickcheck.hh to create you test. Make a ZeroStartPriceGenerator class, with a PriceSimulation member, and a reset function to set all its parameters:

Diffuse/StockPriceTest/PropertyBasedTests.cpp

```cpp
class ZeroStartPriceGenerator
{
public:
  ZeroStartPriceGenerator() : price_(0.0, drift_, 0.0, 0.1) {}

  void reset(double drift, double dt, int sims, unsigned int seed)
  {
    drift_ = drift;
    dt_ = dt;
    sims_ = sims;
    seed_ = seed;
    price_ = Diffuse::PriceSimulation(0.0, drift_, 0.0, dt_, seed);
  }

  double Seed() const { return seed_; }
  double Drift() const { return drift_; }
  double Dt() const { return dt_; }
  int Sims() const { return sims_; }

  std::vector<double> prices() const
  {
    std::vector<double> prices;
    for(int i=0; i<sims_; ++i)
      prices.push_back(price_.Next());
    return prices;
  }

private:
  double drift_;
  double dt_;
  int sims_;
  double seed_;
  mutable Diffuse::PriceSimulation price_;
};
```

Use this in a property ZeroStartPriceGivesZero class to test your simulations:

Diffuse/StockPriceTest/PropertyBasedTests.cpp

```cpp
class ZeroStartPriceGivesZero : public Property<ZeroStartPriceGenerator> {

  bool holdsFor(const ZeroStartPriceGenerator& gen)
  {
    std::vector<double> xs = gen.prices();
```

```
      for(const auto & p : xs)
        if (p != 0.0) return false;
      return true;
    }

    bool accepts(const ZeroStartPriceGenerator& gen)
    {
      return gen.Dt() > 0.0;
    }
};
```

State the property the generated item holdsFor. Get your generator's prices here. A non-zero price (p) is a failure, so return false if that happens. The framework will report the numbers used in the generator if this ever fails. You can turn the specific example into a unit test or fix your problem—or both.

Now create an overload of generate for your class. Use the built-in generators for int and float to choose the numbers for your simulation and use the reset function to set up an example:

Diffuse/StockPriceTest/PropertyBasedTests.cpp

```
void generate(size_t n, ZeroStartPriceGenerator & out)
{
  double drift, dt;
  int sims;
  unsigned int seed;
  generate(n, drift);
  generate(n, dt);
  generate(n, sims);
  generate(n, seed);
  if (dt < 0) dt *= -1;//filter out negatives
  out.reset(drift, dt, sims, seed);
}
```

Declare your property checks in main and you automatically get a set of tests, using the generate function. Choose how many you want to check; this uses 100:

```
int main()
{
  ZeroStartPriceGivesZero zeroStartPrice;
  zeroStartPrice.check(100);
}
```

When you run this you should see OK, passed 100 tests.

You can probably think of other properties to check:

- Stock with zero volatility has an average move of its drift
- The average displacement of a Brownian motion particle is 0
- The variance of displacement of a Brownian motion particle is 1

The property tests may pass if it doesn't happen to pick a bad combination, but enough generated combinations and the so-called *shrinker*, seeking out values to make the property fail, makes this unlikely. The Hypothesis blog has lots of details about shrinkers if you want to read more about them.[11]

Property-based testing is a good complement to unit testing, and worth considering if you are using randomly generated numbers. It's also good for deterministic code. It's worth learning about and trying out.

Over to You

You built three simulations in this chapter, used two different statistical distributions, drew pictures in C++, and briefly looked at property-based testing. The models you simulated used stochastic differential equations—equations with a random element. They were a special type of random walk known as Markov processes; the next step depends on the current state only. You built a discrete simulation of these by finding values at specific time intervals.

Think about some other property-based tests for your code. Look at unit tests in a project you are working on—can you see any magic numbers? Try property-based testing on that code and see what happens.

Try out some different models. Your stock prices tend to go up over time. Interest rate models are often mean reverting—giving you another property to check—they go up and down, but lean to a long-term average. Vasicek is one of several interest rate models. Instead of tending to drift up over time, they tend to oscillate around a mean value. If you search online for "interest rate sde" you will find lots of related models. It uses the SDE

$$dr_t = a(b - r_t)dt + sdW_t$$

This has the familiar dW, and few constants; the speed to reversion to the mean (a), the long term mean (b), and volatility (s). *Options, Futures and Other Derivatives [Hul06]* contains further details of pricing models if you want to read more.

There are many other application areas for stochastic simulations. In general, a Monte Carlo simulation gives you a trial and error way to solve a problem. Some stochastic models take you nearer to probability or statistical learning than machine learning. Though the edges are fuzzy, the building blocks are similar.

---

11. hypothesis.works/articles/integrated-shrinking/

In the next chapter, you will combine the small steps from this chapter with a fitness function to make a bee swarm. The bees will explore and eventually buzz out of a paper bag en masse. This is another swarm intelligence algorithm, so it builds on what you have already learned. The bees will converge on one solution, and you will revisit the ideas of tournament and roulette wheel selection from genetic algorithms, and the co-operation of previous swarm algorithms to keep local and global sweet spots in the play.

CHAPTER 7

Buzz! Converge on One Solution

In the previous chapter, you simulated diffusion, drew the results in C++, and considered how to test code with random elements. Your particles eventually dispersed or diffused as intended. They took a long time to escape the bag—unless you cheated with the input parameters—even going as far as turning off the stochastic behavior for the stock price simulation.

If you're wondering how to make everything escape quickly, your earlier foray into particle swarms and ant colonies laid the foundation of the general idea of swarm optimizations or intelligence (SI). In this chapter, you'll add to your knowledge by building another swarm optimization—this time, bees.

Imagine you've got some bees buzzing around in a paper bag looking for food. After exploring, the bees return home. But what happens if the bees find a better source of food outside of the bag? With this particular swarm, the bees will abandon their hive quickly and flee en masse to a new home. Of course, real bees tend to swarm when extra queens are hatched, but these nature-inspired abstract bees can do whatever they want.

Your algorithm will have the familiar "for a while" loop, which you can stop after all of the bees are out of the bag. Your ants left a pheromone trail to communicate. Your bees will also have *stigmergy*—a term borrowed from biology to indicate agents building a consensus by communicating. Your bees will do a *waggle dance* to communicate back at their hive after exploring.

With this example, you'll see how easy it is to take the bare bones of an algorithm and tweak it to your own ends. You'll rediscover ways to use local and global information as you solve a problem. You'll code an *Abstract Bee Colony* (ABC) and think about options and alternatives.

Typical outlines of a machine learning algorithms leave you guessing exactly how to implement them. You almost always have to choose parameters and need to get your algorithm started. If this isn't made clear, don't panic. Try something and see what happens. You've already had practice making a decision when the exact details of a recipe are not immediately apparent. You'll have decisions to make in this chapter too.

Your Mission: Beekeeping

Your bees will have different roles. Some bees will wait at home, some will scout around exploring, and other bees will work to collect pollen from a food source. You can plant food at specific points in space, or put food anywhere to keep things simple. Pretend you dropped off better food near the top of the bag and the best food outside of the bag. You'll encode the quality of a food source as a fitness function, encouraging bees to find better food sources. You get an *optimal* solution to a problem when the bees find the best food source. If better food is higher up and the best food is outside the bag, your bees will learn to move up, eventually escaping the paper bag. The position of a food source encodes a solution to your problem. You'll end up with a position (x, y) that is outside of the bag. By now, the elements of optimization, fitness function, and global and local searches should all be familiar to you.

Get Your Bees Buzzing

You can start by choosing a single food source for the bees somewhere near the bottom right side of the bag. You can have more than one known food source to begin with, however, you don't need to have more than one. When in doubt, start simple. You can also experiment where to place the food. If it's nearer to the top of the bag, the bees won't take as long to escape. You also need a place for their hive. You can pick anywhere in the bag, say the bottom-middle. The following figure shows your bees ready to start exploring:

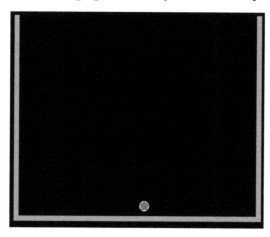

Your bees have a home and some (invisible) food to seek out. The bees will have one of three roles, each defining how they explore. You can represent the different roles with different shapes and colors. Over time, the bees will find more food, cooperate, and learn how to get out of the bag.

The Many Roles of Bees

The first food source represents local knowledge. *Worker* bees will make a beeline here, pun intended, and buzz around slightly on the way, then return home. The next image shows a potential route for a worker bee. The path combines a buzz right and up, to give a beeline toward the food, nudged by a small random jiggle on the way:

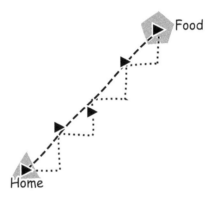

Meanwhile, *scout* bees will explore the wider space and freely buzz around anywhere. These bees represent the global part of the search. The scout bee movements are similar to the particle swarm and diffusion model, making random steps. They'll prefer better places. A scout bee is not interested in the previous food source. It remembers this, but its mission is to explore, trying to find better food. The next image shows a possible route for a scout bee. It buzzes around exploring but tends to move upward:

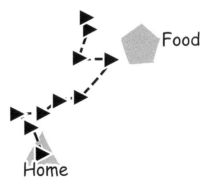

Other bees, *inactive* for now, wait at home. They remember where they found food before and wait for the others to return home.

Your bees perform their roles together. The worker bees make a beeline for the food you initially told them about in the right-hand corner, as you can see in the next image. The scouts explore elsewhere and tend to go up. The inactive bees wait at home. You'll see how to draw the bees and make them move later when you make bees swarm on page 133. For now, the image shows what happens when your bees begin to learn. The left image shows them starting to venture out, and the right image shows the worker bees near the first food source, while the scout bees explore elsewhere:

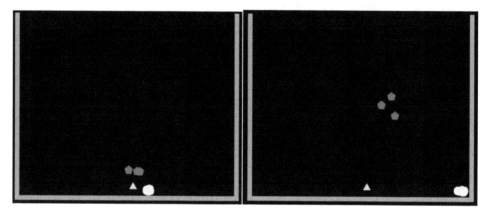

Those that venture out eventually return home and communicate to the others via a waggle dance that indicates the best food they found. You don't need to code the waggle dance. Your bees will tell each other the best food spots and move from side-to-side to indicate the dance.

Joe asks:
What's a Waggle Dance?

When a bee returns to the hive, it waggles in a figure eight pattern. The direction of the figure eight points to the food, and the length of the waggle indicates the distance. The angle is relative to the sun, and the bee is clever enough to adjust the angle of its dance as the sun moves. The better the location, the faster the bee waggles, getting attention from others.

The waggle dance happens in the hive (hexagon) and points at the food (small circle) in the next picture. The length of the dance indicates the distance to the food. The bee will change the angle of the dance as the sun (big circle) moves. Something like you see in the following image—although the dance happens in the hive, so this isn't to scale:

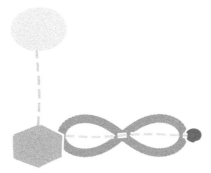

The watching bees consider updating their favored place to find nectar. The bees will remember their best food source so far, so you need a fitness function to decide which is best. The bees gradually find better food sources, guided by the fitness function. The entire process repeats for a while, stopping when the bees swarm out of the bag.

Overview of ABC

The bees swarm when they find a new food source outside the bag. Given a continuous two-dimensional space, getting two bees to agree on the same spot is difficult. A whole bee colony opting for the same single point in space is even harder. So you can cheat. Slightly. You can check bees' food sources when they get home. If all of these are outside the bag, choose the best food source for simplicity. This is enough to get an overview of how the algorithm works. You're trying to get the bees out of the paper bag, so any position outside will do.

How to Feed the Bees

Pulling together the algorithm gives you something like this:

```
For a while
  Go out
    Worker bees
      get food from a known food source
        and explore nearby
    Inactive bees
      wait at home
    Scout bees
      explore remembering the better food sources
  Go Home
  Waggle dance
    recruit bees
  Maybe swarm
```

You need different types of bees, controlling how they explore. You can draw them as differently colored polygons when you code this so you can tell which is which.

Decisions to Make

You need to fill-in some blanks to make your bees swarm. Let's start with a few questions and some possible answers.

1. Where should the bees start?

2. Where should you put food for worker bees?

3. How do they decide their favorite spot?

4. What proportions of bees should you assign to each role?

5. How do you assign the roles?

6. How many bees for each role?

7. How do you define the fitness function?

8. What happens in the waggle dance?

For this exercise, you're going to start the bees in the bottom of the bag. If you start your bees too near the top for this recipe, they'll buzz out of the bag very quickly. Plus, by starting near the bottom, you give them a bit more of a challenge. When faced with real-world problems, you can make similar choices or choose a starting point at random.

Tell the bees there's food in the bottom-right of the bag to begin with. Any other spot they explore will have some food. For some problems, only certain points or solutions are worth considering, but for this chapter, you'll allow any point. It makes no difference for this problem and means you don't need to keep a list of possible points. If you tried the traveling salesman problem instead, which you saw on page 79, you would need to restrict the points visited. The fitness function will weed out worse places. The bees use this to decide their favorite spot. Higher up will be better as it encourages the bees to travel upward. You therefore compare the y coordinate of two positions to decide the best.

You'll give each bee an enumeration value to dictate its role. Select the number of workers, scouts, and inactive bees you want, and send these into main making it easy to try different ratios. If all of the bees are inactive, no exploration will happen. If all of the bees are workers, they'll take a long time to get out of the paper bag, since they only explore a little. Therefore, it makes sense to have at least three bees, one for each role.

You'll make a Hive class to control the bee colony and an update function will drive their exploration. You need to decide for how many steps they explore. When these are up, tell the bees to go_home. If the bees all find food outside of the bag, you'll tell them to swarm. Let's see how they communicate using the waggle dance when they go_home, then you have all you need to make bees swarm.

Waggle Dance

You can draw the bees moving side-to-side to indicate the waggle dance but code the information sharing separately.

One way to implement this is by randomly selecting two bees. Swap their roles, and compare notes on the best food source using your fitness function. The bees then update their favorite food source, remembering it for the next outing. Worker bees may find slightly better food near their favorite place. Scout bees have explored further, so may report much better places. Inactive bees remember what happened last time they explored. In each case, the better place gets passed on. The inactive bees retain their memory of what happened before just in case the current exploration doesn't go as well. This gives your bees some long-term memory.

Instead of always selecting a better food source, you can make a probabilistic decision which you saw in the genetic algorithm chapter on page 39. For this, you can use a roulette wheel or tournament selection. Keep in mind, always going for the better option now means you might miss something even better later on, which may or may not matter to you. In some cases, you may need some experimentation or awareness of the complexity of your search space. However, with these bees, they'll swarm if you choose the best each time, simplifying the implementation. Ready for the code?

Let's Make Some Bees Swarm

Make a new C++ project and add SFML—which you saw on page 110—so you can draw the bees buzzing. The code with this book has a Bees static library for the algorithm. This is called by a unit test project and a main project named ABC. Make the library first, and feel free to add tests as you go. The tests aren't shown here.

Code Your ABC

Start by defining a Coordinate class:

Buzz/Bees/Bee.h
```
struct Coordinate
{
  double x;
  double y;
};
```

You need a Bee class with a home, current position, a favored food spot, and a role.
They start out at home. Add a buzz to drive how far the bees move each time.
You'll fill in the remaining parts to this shortly:

Buzz/Bees/Bee.h
```
class Bee
{
public:
  explicit Bee(Role role,
    Coordinate position = { 0.0, 0.0 },
    Coordinate food = { 0.0, 0.0 },
    double buzz = 5.0)
    : role(role),
    position(position),
    home(position),
    buzz(buzz),
    food(food)
  {
  }

  Role get_role() const { return role; }

  void communicate(Role new_role, Coordinate new_food);
  void scout(double x_move, double y_move);
  void work(double x_move, double y_move);
  void go_home();

  bool is_home() const
  {
    return (position.y > home.y - buzz)
        && (position.y < home.y + buzz)
        && (position.x > home.x - buzz)
        && (position.x < home.x + buzz);
  }

  void waggle(double jiggle);
  Coordinate get_pos() const { return position; }
  Coordinate get_food() const { return food; }
```

```
  void move_home(Coordinate new_home)
  {
    home = new_home;
  }
private:
  Role role;
  Coordinate position;
  Coordinate home;
  Coordinate food;
  const double buzz;
};
```

This doesn't do much yet. You can find where a bee currently is using get_pos, and you can see what role it has using get_role. You can see if it is_home, give or take a small buzz. You can also make it move_home, which you'll use when your bees swarm. First things first, though.

You need to create a Hive for your bees. This Hive updates your bees, tells them when to come home, and tells you if they are all_home. It also makes them swarm:

Buzz/Bees/Bee.h
```
class Hive
{
public:
  Hive(int number_workers,
    int number_inactive,
    int number_scout,
    Coordinate start_pos, Coordinate food, float buzz, int steps);

  std::vector<BeeColony::Bee> get_bees() const
  {
    return bees;
  }

  void update_bees();
  void swarm();
  bool all_home();

private:
  std::vector<Bee> bees;
  const size_t steps;
  size_t step;
  std::mt19937 engine;
  std::normal_distribution<double> normal_dist;
  std::uniform_int_distribution<> uniform_dist;

  void waggle_dance();
  void explore();
};
```

You use the normal_dist for the bees' movement. You use the uniform_dist to choose which bees swap roles and exchange information on food sources. You then initialize it with the total number of bees. Each number drawn corresponds to a distinct bee, and each bee is equally likely to be picked. Create enough bees in the constructor, telling them where to start, where their food is initially, and how many steps to explore for:

Buzz/Bees/Bee.cpp

```
Hive::Hive(int number_workers,
    int number_inactive,
    int number_scout,
    Coordinate start_pos, Coordinate food, float buzz, int steps)
    : bees(bees), steps(steps), step(0u),
    engine(std::random_device()()),
    uniform_dist(0, number_workers+ number_inactive+ number_scout - 1)
{
  for (int i = 0; i < number_inactive; ++i)
  {
    bees.emplace_back(Role::Inactive, start_pos, food, buzz);
  }
  for(int i=0; i < number_workers; ++i)
  {
    bees.emplace_back(Role::Worker, start_pos, food, buzz);
  }
  for(int i=0; i < number_scout; ++i)
  {
    bees.emplace_back(Role::Scout, start_pos, food, buzz);
  }
}
```

You update_bees in the Hive, moving them out or home. The role dictates how they move. Remember, inactive bees will wait at home, while the Worker and Scout bees need to move. When a Bee moves, it may notice better quality food, and it remembers the new place accordingly which means you need a fitness function. Use the food's Coordinate to decide how good a spot is. For this problem, the height (y) is all you need:

Buzz/Bees/Bee.h

```
inline double quality(Coordinate position)
{
  return position.y;
}
```

Your bees can now work out which is the best spot so far.

The bees are either moving about, coming home, waggling (just for visual effect) or performing their waggle_dance. Code this in your update_bees function:

Buzz/Bees/Bee.cpp

```cpp
void Hive::update_bees()
{
  //either moving or waggling then updating
  if (++step < steps)
  {
    explore();
  }
  else if(!all_home())
  {
    for(auto & bee : bees)
    {
      if (!bee.is_home())
        bee.go_home();
      else
        bee.waggle(normal_dist(engine));
    }
  }
  else
  {
    waggle_dance();
  }
}
```

Let's work through the code. First, your bees explore for a few steps:

Buzz/Bees/Bee.cpp

```cpp
void Hive::explore()
{
  for (auto & bee : bees)
  {
    switch (bee.get_role())
    {
    case Role::Worker:
      bee.work(normal_dist(engine), normal_dist(engine));
      break;
    case Role::Scout:
      bee.scout(normal_dist(engine), normal_dist(engine));
      break;
    }
  }
}
```

Generate a move with the normal_dist via your engine. If you aren't accustomed to this way of generating random numbers, look back to the example of rolling a die on page 110. You need horizontal and vertical moves for the scout or worker bees. The default distribution gives you numbers centered around 0, with a variance of 1. This means about two thirds are likely to be between -1 and +1, while 95% are likely to be between -2 and +2.[1] Use the buzz you set in the Bee's constructor to scale up this random number, making the moves as large as you like. The scout only moves to better quality positions:

Buzz/Bees/Bee.cpp
```cpp
void Bee::scout(double x_move, double y_move)
{
  Coordinate new_pos{position.x + buzz * x_move, position.y + buzz * y_move};
  double new_quality = quality(new_pos);
  if (new_quality > quality(position))
  {
    food = new_pos;
    position = new_pos;
  }
}
```

Don't forget, you can make this probabilistic instead, allowing slightly worse updates from time to time.

A worker will move approximately half the step size. This bee will tend toward its favored food position but jiggle around slightly. Step across and up (or down) toward the food, as you saw in the worker moving sketch on page 129. Then add the jiggles, x_move and y_move, to the position conditional on the quality:

Buzz/Bees/Bee.cpp
```cpp
void BeeColony::move(Coordinate & from, const Coordinate to, double step)
{
  if (from.y > to.y)
    from.y -= step;
  if (from.y < to.y)
    from.y += step;
  if (from.x < to.x)
    from.x += step;
  if (from.x > to.x)
    from.x -= step;
}

void Bee::work(double x_move, double y_move)
{
  move(position, food, buzz/2.0);
  double new_quality =
    quality({ position.x + x_move, position.y + y_move });
  if (new_quality >= quality(position))
  {
    position.x += x_move;
    position.y += y_move;
  }
}
```

---

1.   en.wikipedia.org/wiki/68%E2%80%9395%E2%80%9399.7_rule

Notice the scout might update its food while the worker is only "collecting pollen." You can update a worker's best position too if you want. Inactive bees wait at home, so you don't need to code anything for their moves.

You keep track of the exploration time using step in the update function on page 137. Tell the bee Hive to go_home when they have tried enough steps. Send the bees home one step at a time. This move is identical to bees going toward the food, but with a different target:

Buzz/Bees/Bee.cpp
```
void Bee::go_home()
{
  if (!is_home())
  {
    move(position, home, buzz);
  }
}
```

Make the bees waggle as follows:

Buzz/Bees/Bee.cpp
```
void Bee::waggle(double jiggle)
{
  if (get_role() == Role::Inactive)
    return;

  position.x += jiggle;
}
```

Your bee moves from side to side while waiting for the others to get home. Once all the bees are home, they share information on the best food sources in waggle_dance. For each bee, find another to communicate with and then reset the step count as follows:

Buzz/Bees/Bee.cpp
```
void Hive::waggle_dance()
{
  for (auto & bee : bees)
  {
    const size_t choice = uniform_dist(engine);
    const auto new_role = bees[choice].get_role();
    const auto new_food = bees[choice].get_food();
    bees[choice].communicate(bee.get_role(), bee.get_food());
    bee.communicate(new_role, new_food);
  }
  step = 0;
}
```

The bees swap roles and decide the best food source when they communicate:

Buzz/Bees/Bee.cpp
```cpp
void Bee::communicate(Role new_role, Coordinate new_food)
{
  role = new_role;
  if (quality(new_food) > quality(food))
  {
    food = new_food;
  }
}
```

By swapping bees' roles, you keep the same number of scout, worker, and inactive instances over the life of the algorithm. For a real-world problem, you might need to vary this. If you're struggling to find a solution, try more scouts to allow more exploration. You can also add more inactive bees to remember more possible solutions in your food data. You can persuade a bee to change roles instead of both swapping to vary the proportions as your ABC runs.

You need one final piece and your ABC is complete. When all the bees have a food source above the bag, the bees will swarm. Check the quality of each bee's food source against this target:

Buzz/Bees/Bee.cpp
```cpp
bool BeeColony::should_swarm(
    const std::vector<BeeColony::Bee> & bees,
    double target)
{
  return bees.end() == std::find_if(bees.begin(), bees.end(),
    [target](const Bee & bee) {
      return quality(bee.get_food()) < target;
  });
}
```

If you decide they should_swarm, make your bees swarm to the best food source:

Buzz/Bees/Bee.cpp
```cpp
void Hive::swarm()
{
  double best_x = -1.0, best_y = -1.0;
  for(const auto & bee : bees)
  {
    if(quality(bee.get_food()) > best_y)
    {
      best_y = bee.get_food().y;
      best_x = bee.get_food().x;
    }
  }
```

```
  for(auto & bee : bees)
  {
    bee.move_home({ best_x, best_y });
  }
  step = steps;
}
```

Set the step to the maximum value to indicate that exploration is over. You now have a complete abstract bee colony.

Display Your ABC

Make your hive and draw the bees in action. Draw the bag, as you've done before on page 115 and call update while the SFML window is open:

Buzz/ABC/main.cpp
```
void action(BeeColony::Hive hive,
  float width,
  float edge,
  float bee_size = 10.0f)
{
  const float lineWidth = 10.0f;
  const float height = 400.0f;
  const auto bagColor = sf::Color(180, 120, 60);
  sf::RenderWindow window(
    sf::VideoMode(
      static_cast<int>(width + 2*edge),
      static_cast<int>(height + 2*edge)
    ),
    "ABC");

  bool paused = false;
  bool swarmed = false;
  while (window.isOpen())
  {
    sf::Event event;
    while (window.pollEvent(event))
    {
      if (event.type == sf::Event::Closed)
        window.close();
      if (event.type == sf::Event::KeyPressed)
        paused = !paused;
    }

    window.clear();
    draw_bag(window, lineWidth, edge, height, width, bagColor);

    if (!paused)
    {
      hive.update_bees();
```

```
    if (!swarmed && should_swarm(hive.get_bees(), height + bee_size))
    {
      hive.swarm();
      swarmed = true;
    }
  }
  draw_bees(hive.get_bees(), window, bee_size, edge, width, height + edge);
  window.display();

  std::this_thread::sleep_for(std::chrono::milliseconds(50));
  }
}
```

Draw your bees making sure they don't go through the sides. Your ACO does not constrain the bees, so do it now. This is just for display, but you can filter out failed solutions to real-world problems in a similar manner:

Buzz/ABC/main.cpp
```
void draw_bees(const std::vector<BeeColony::Bee> & bees,
  sf::RenderWindow & window,
  float size,
  float edge,
  float width,
  float height)
{
  for(const auto & bee : bees)
  {
    sf::CircleShape shape = bee_shape(size, bee.get_role());

    float x = static_cast<float>(edge + size + bee.get_pos().x);
    if (x > edge + width - 2*size)
      x = edge + width - 2*size;
    if (x < edge + 2*size)
      x = edge + 2*size;
    float y = height - 2 * size - static_cast<float>(bee.get_pos().y);
    shape.setPosition(x, y);
    window.draw(shape);
  }
}
```

Distinguish the bees by role, using differently colored polygons. The simplest way to draw a polygon in SFML is to create a CircleShape and setPointCount. By default a CircleShape has 20 points—enough to look like a circle—and you can choose different values to make regular polygons:

Buzz/ABC/main.cpp
```
sf::CircleShape bee_shape(float size, BeeColony::Role role)
{
  sf::CircleShape shape(size);
  switch (role)
```

```
  {
    case BeeColony::Role::Worker:
    {
      shape.setPointCount(20);
      shape.setFillColor(sf::Color::Yellow);
    }
    break;
    case BeeColony::Role::Inactive:
    {
      shape.setPointCount(3);
      shape.setFillColor(sf::Color::Cyan);
    }
    break;
    case BeeColony::Role::Scout:
    {
      shape.setPointCount(5);
      shape.setFillColor(sf::Color::Magenta);
    }
    break;
  }
  return shape;
}
```

Call action from main in your ABC console application. Try different proportions of the bee roles. See what happens.

Did It Work?

You had several choices to make for this algorithm. You started the bees and the first food source in a fixed place. You then had to decide how many bees had each role. The code in this book uses ten worker bees, five inactives, and three scouts by default. Your choice affects the number of updates before the bees swarm out of the bag.

You can reason about what happens for some setups. If you have:

- A single bee which is inactive, nothing will ever happen.

- A single worker bee, and no others, it gathers food from one spot and only tries nearby spots, so takes a long time to move up.

- A single scout bee, it can manage to escape from the bag, though it doesn't do any machine learning.

For greater numbers of bees, run an experiment to see what happens. The bees tend to swarm out of the bag after about 600 updates when you use the default ratios. If you have one bee in each role, on average, they only need 400 or so updates. However, this sometimes takes much longer. If you have five of each, they need an average of 500 iterations until they swarm. Compared to a single bee in each role, this is more consistent, as though they are working as a proper bee colony. You can explore to find a cut-off point where they behave more like a colony communicating than individuals.

If you use reasonable proportions, the bees all swarm out of the bag eventually. You did cheat somewhat to ensure this happened. You made your bees pick higher spots, so you forced them to go up over time. The precise paths taken vary each time, but the general behavior is the same. Your worker bees find more food sources quite quickly, as shown in the following figure. The yellow circle worker bees are now going to two food sources—the original one in the bottom-right, and a new one higher up:

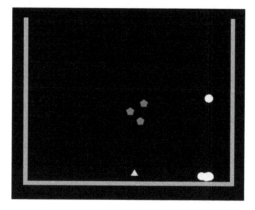

Gradually more food sources are located, higher up each time. Eventually, they all have food sources outside of the bag. When this happens, your bees swarm to one of these, as shown in the following image:

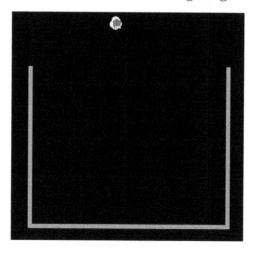

Excellent beekeeping. You helped your bees find a food source outside of the paper bag.

Over to You

You now have a working ABC. You saw the global search from the scout bees and the local search from the worker bees. You combined these to get all of the bees out of your paper bag. You can apply this to a variety of real-world applications, from training neural networks, to improving the performance of automatic voltage regulator systems, and even for clustering and feature selection in data mining.[2] Furthermore, several articles have been written about using a bee colony for testing software:

- Automated Software Testing for Application Maintenance by using Bee Colony Optimization algorithms (BCO) by K. Karnavel and J. Santhoshkumar 2013[3]

- An Approach in the Software Testing Environment using Artificial Bee Colony (ABC) Optimization by T. Singh and M. K. Sandhu 2012[4]

- Testing Software Using Swarm Intelligence: A Bee Colony Optimization Approach[5]

James McCaffrey provides a walkthrough of this algorithm for the traveling salesman problem.[6] He calls this a "simulated bee colony." Watch out for the different variations of the name.

There are many different nature-inspired swarm algorithms, including:

- Glowworms
- Cats
- Roach infestations
- Fish schools
- Leap frogs (I'm not joking)[7]

You might even be able to invent your own now that you've learned about bees, particles, and ants.

---

2. dl.acm.org/citation.cfm?id=2629886
3. http://ieeexplore.ieee.org/document/6508211/
4. pdfs.semanticscholar.org/e6dc/153350972be025dc861fc86e495054e85d37.pdf
5. dl.acm.org/citation.cfm?id=2954770
6. msdn.microsoft.com/magazine/gg983491
7. arxiv.org/pdf/1307.4186.pdf

You have implemented several swarm algorithms and used many fitness functions. You have used an up-front model to see what happened when you made Monte Carlo simulations. You can also create some up-front rules to drive interactions between agents or cells. The cells automatically respond to the state of their neighbors, giving rise to automata.

In the next chapter, you'll create cellular automata. You'll start with a random initial population, as usual, but you'll have rules governing whether a specific cell lives or dies. You'll put some cells in a paper bag. As the state of your cells change, patterns can emerge. Sometimes static patterns form; sometimes a pattern oscillates or cycles through states. Once in a while, a pattern seems to glide up, so it might get out of the paper bag over time. This draws a line under swarm algorithms. Cellular automata have a very different feel. They stray closer to *artificial intelligence* than machine learning, but you'll find a good starting setup using a genetic algorithm later on page 163. You need to build a simple cellular automaton first. You'll see complex behavior emerges from some simple rules. If you get lucky, you might end up with some live cells outside your paper bag.

CHAPTER 8

Alive! Create Artificial Life

In the previous chapter, you made an abstract bee colony and used a fitness function to guide the bees' movement. The bees remembered the best places they had visited, and they communicated these places with the other bees by performing a waggle dance. This encoded information-sharing between agents, encouraging them to learn. In the end, the bees swarmed out of the paper bag when they all found food outside.

Now, imagine cells on a grid, some of which are in a paper bag. If a cell gets crowded, it dies; however, when certain conditions are met, a cell remains alive or even comes to life. These live cells can form a stable shape, or cycle through states, making patterns. You'll see this as you play with Conway's *Game of Life* throughout this chapter and end up with live cells outside of your paper bag.

The idea dates back to the 1940s and was publicized by Martin Gardner in *Scientific American* in 1970. The original version investigated a theoretical machine that could copy itself, hypothesizing large-scale mining of asteroid belts by self-replicating spaceships. Some early artificial intelligence ideas stray into the realm of science fiction.

This rule-based approach differs from the previous algorithms since you don't have a model, a target to achieve via fitness functions, or a random heuristic search. Instead, you follow a set of rules governing whether points or cells are dead or alive. The rules form a *cellular automaton* (CA). All CA are governed by a simple set of rules leading to *emergent* behavior. Many are universal Turing machines or *Turing complete*. You can, therefore, use them to code programs, but that's beyond the scope of this chapter.

You can use cellular automata to solve real-world problems. You made a classifier in Chapter 2, *Decide! Find the Paper Bag*, on page 15, to predict the

Joe asks:
What's Turing Complete?

Alan Turing, an English mathematician, is regarded as the founder of computer science. Turing designed a theoretical machine to investigate the *Entscheidungsproblem* or decision problem—can you create a process to decide whether a mathematical statement is provable or not? See *The Annotated Turing [Pet08]* for more details. The *Turing machine* performs limited operations using symbols on paper tape.

Meanwhile, Alonzo Church, an American mathematician who invented lambda calculus familiar to functional programmers, showed the decision problem is undecidable. Their combined ideas give the *Church-Turing Thesis*—a function on natural numbers can be computed if and only if a Turing machine can compute it. Any system, either a programming language with possibly infinite memory or abstract system, such as lambda calculus, capable of simulating a Turing machine is *Turing complete*.

class of new data. Different classifiers can draw different conclusions. Cellular automata have been used to make a voting system combining the output from several classifiers to make a group decision.[1] They can also be used to create music.[2]

In this chapter, you'll build one CA. You'll see two more in the next chapter and use a genetic algorithm to choose starting states to achieve a goal. Starting with the Game of Life will give you a clear idea of how CA generally work, and you'll see a variety of patterns emerge. One of these, the *glider*, is sometimes called a universal hacker emblem:

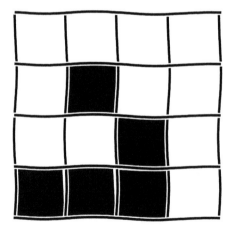

---

1.   www.researchgate.net/publication/221460875_Machine-Learning_with_Cellular_Automata
2.   www.ibm.com/developerworks/library/j-camusic/

A hacker has "technical adeptness and a delight in solving problems and overcoming limits."[3] Hackers should also be able to program their way out of a paper bag. You certainly can by now.

Your Game of Life will take place on grid squares, which will contain immovable cells. The cells are either alive or dead. There are two ways in which a cell can die: isolation and overcrowding. When the conditions are just right, a cell remains alive or comes to life.

You know the story of Goldilocks, sitting in a bear's chair, eating a bear's porridge, and sleeping in a bear's bed, provided it was just right. The fairy story was originally about intruders and burglars, but now people focus on the "just right" aspect. The planet Earth is in just the right place for life to emerge. This circumstellar habitable zone gets called a Goldilocks zone.[4] CA relate to *artificial life* research, investigating how to find sweet spots that might let sustainable life emerge. Artificial intelligence covers big topics, and machine learning is just a small part.

Your Mission: Make Cells Come Alive

There are many named cellular automata. You'll see *elementary cellular automata* in the next chapter. These operate on one-dimensional rows. You can build CAs in two or more dimensions. Christopher Langton, an artificial life researcher, developed a two-dimensional automaton. Langton's CA has an artificial ant on a grid. The ant walks around, coloring squares as it moves. The ant moves forward to one of the four neighboring squares: left, right, up, or down. Two rules govern the ant's behavior:

- On a white square, turn clockwise.
- On a black square, turn counter-clockwise.

In either case, it flips the color of the current square and steps forward.

You tend to see three things happen in this CA. First, the ant makes small, simple patterns, like squares or other symmetric shapes. After a while, the ant devolves into chaos, making a bit of a mess with no obvious pattern. Finally, it builds a *highway* pattern—a straight line made of several black cells moving away from the chaotic mess. No one has proved the highway will always get built, but for any setup tried so far it has been. Could you tell this highway would emerge from these two rules? Probably not. Exploring emergent behavior can be fascinating.

---

3. www.catb.org/hacker-emblem/
4. en.wikipedia.org/wiki/Circumstellar_habitable_zone

You've already made ants crawl out of your paper bag. Feel free to try out Langton's ant and other CAs too. In the meantime, try the Game of Life here. You'll see lots of different patterns emerge, in contrast to the ant's highways.

The Game of Life has four rules:

1. A cell with fewer than two live neighbors dies.

2. A cell with two or three live neighbors lives.

3. A cell with more than three live neighbors dies.

4. A dead cell with exactly three live neighbors comes to life.

In tabular form, these rules boil down to counting neighbors, depending on the current state:

Current state	Cell's live neighbors	New state
Alive	< 2	Dead
Alive	= 2 or = 3	Live
Alive	> 3	Dead
Dead	= 3	Live

Of course, you don't get to play the game; you only get to sit and watch. The artificial life emerges with no further intervention. If you want a way to interact, you can extend this example by making a mouse click bring a cell to life.

You can predict a few patterns that emerge, but no-one has worked out all the things that can happen yet. Let's think about a few simple cases. Each cell has eight possible neighbors, just like your ant did on page 83. A cell needs three neighbors to come to life and two or three live neighbors to remain alive. This means if no cells are alive, no cells can come to life. If only one or two cells are alive, they also die off. You need at least three neighboring cells alive for patterns to form. The patterns can be stable, cycle through states, or move across the grid.

What happens with a block of four live cells—two by two? Not much. Each live cell has three live neighbors, and the others have no more than two. The live cells stay alive, and no new cells come to life, so the block stays where it is forever. There are other stable patterns too. The block of four is the simplest.

How do you get a pattern that changes, cycling through states? Think about what happens when you have three live cells in a line. Count how many live neighbors surround each cell this time as shown in the figure on page 151.

Initially...	Then...	Finally...

2	**1**	2
3	**2**	3
2	**1**	2
1	1	1

1	3	1	1
1	**2**	**1**	1
1	3	1	1

The central live cell always has two live neighbors, so it stays alive throughout. The other live cells have this central cell as the only live neighbor, so they die off. Two of the empty cells have three live neighbors, so they come alive. The column of three cells, therefore, transforms into a row of three cells. This row then turns back into a column of three cells. This type of pattern is an *oscillator* known as a *blinker*. Your blinker swaps between two patterns, so it has a *period* of two. These cycling patterns stay in the same place. A *spaceship* also cycles through states, but it moves as it changes. A common spaceship is a *glider* you saw on page 148. You'll see a variety of patterns when you assess your CA.

Your Game of Life will take place on a fixed-sized grid with the usual rules. You can extend or alter the rules by changing how many need to be alive or which cells are the neighbors. You can wrap the grid to form a cylinder or even a torus (donut). To make a cylinder, make the top spill round to the bottom, and vice versa. To make the torus, join the ends of your cylinder:

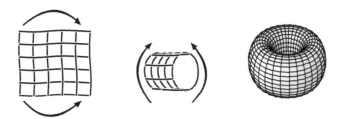

You now have the general idea of how this specific CA works and some types of patterns that can form. You have a few decisions to make before you can implement this. In the next section, you will think about your grid size, how to represent and update your cells, and how to find the neighbors. After deciding, you'll implement this in C++.

How to Create Artificial Life

The overall algorithm for the Game of Life is simple:

```
grid = setup()
forever:
  new_grid = []
  for cell in grid:
    new_grid.push(rules.apply(cell))
    grid = new_grid
```

You update each cell according to the rules, for as long as you want. Previously, you had a choice about when to make updates for ants, bees, and other agents. You still have that choice in a CA, however, a *batch update* is canonical for the Game of Life. You will, therefore, update your cells *offline* by making a new grid based on the rules and current grid state.

You can use asynchronous or *online* updates in CAs instead, by updating the current grid. You can even update the nearest neighbors of a cell rather than the whole grid. These extensions are an active area of research. Blok and Bergersen published a detailed evaluation of the differences between the standard approach and updating a few cells together.[5] They found the number of cells alive initially and the number of cells updated together impacted the outcomes, though a frozen set of patterns tended to emerge. Various approaches to updates have been considered, including randomly choosing a cell to update.[6]

Many machine learning or AI algorithms have a standard or original form. By wondering what else you can do, you can develop some novel research. Before breaking the rules, however, it's useful to understand them, so let's get back to the canonical Game of Life.

Decisions to Make

Even with this simple algorithm, you still need to make some decisions:

- How big is the grid?
- Will you wrap the edges round to make a torus or not?
- Which cells are alive to begin with?
- How will you store the current state of the cells?

The first three choices impact how many patterns you'll see. When you can't decide what to do, set up parameters so you can experiment. Try a relatively

---

5.   www.researchgate.net/publication/235499696_Synchronous_versus_asynchronous_updating_in_the_game_of_Life

6.   en.wikipedia.org/wiki/Asynchronous_cellular_automaton

large grid of 40x50, with a 40x40 paper bag. This choice is big enough for a few patterns to emerge. The extra 10 units leave spaces for cells to come alive outside of the bag—and if you wrap the grid into a torus, the patterns can move around more. Try different sizes—and different shapes—by changing the neighbor-finding function.

Which cells are alive to start with? You want at least three in a line. Otherwise, nothing will happen. If you can't decide which, randomly give life to about half of the cells inside of the paper bag. Again, you can make this configurable, either starting with specific cells alive, or letting your algorithm choose.

To store the state, you can use an std::vector, with one item per cell. Make each item a bool to store a cell's state. Herb Sutter warns this is not a container.[7] Howard Hinnant says it doesn't play nicely with range-based for loops.[8] However, it does provide a quick way to dynamically choose the size of the grid. You can use the Boost library's dynamic_bitset if you can't bring yourself to use an std::vector of bool.[9]

Whichever storage you use, you need to switch between a cell's (x, y) coordinate and its index. Label your grid from 0 in the bottom left, working across a row for the width. Stack up rows as you go. Notice the index is y multiples of the width plus the x value, as the next picture illustrates:

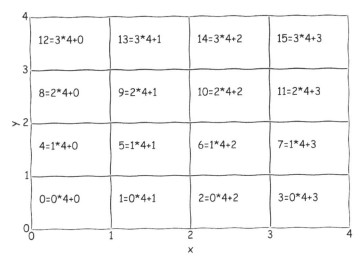

In code you need:

```
size_t index = y*width + x;
```

7.   http://www.gotw.ca/gotw/050.htm
8.   isocpp.org/blog/2012/11/on-vectorbool
9.   http://www.boost.org/doc/libs/1_65_1/libs/dynamic_bitset/dynamic_bitset.html

When you need to go the other way, getting a coordinate from an index, first find how many rows you've filled to get y. To do this, divide the index by the row width. What remains tells you how far along the current row you are. Use the modulus operator % to find this x value:

```
size_t y = index / width;
size_t x = index % width;
```

You can then use the index to store and retrieve the current state of your cells in your std::vector. As your artificial life emerges, you can use the SFML, which you first saw on page 110, to display the current state. Time to code.

Let's Make Cellular Automata

Your cells live on a grid with a fixed Height and Width. You'll need to know which cells are Alive, and you'll want to Update them together, so make a World class to hold your grid and Update your cells:

Alive/GameOfLife/GoL.h
```cpp
class World
{
public:
  World(size_t max_x, size_t max_y, bool wrap);
  World(size_t max_x, size_t max_y, bool wrap,
        size_t start_width, size_t start_height,
        size_t number);
  size_t Width() const { return max_x; }
  size_t Height() const { return max_y; }
  size_t Alive() const;

  bool Alive(size_t x, size_t y) const
  {
    return state[y*max_x + x];
  }

  void Spark(size_t x, size_t y)
  {
    if(Alive(x,y))
      throw std::invalid_argument("Cell already alive");
    state[y*max_x + x] = true;
  }

  void Update();

private:
  const size_t max_x;
  const size_t max_y;
  std::vector<bool> state;//evil
  const bool wrap;

  bool StayAlive(size_t x, size_t y) const;
};
```

You need a way to make some cells come to life. Otherwise, nothing will happen. You can do this in two ways:

- With a constructor taking a number of cells to bring to life.
- With a constructor bringing no cells to life, and a Spark method to bring a specific cell to life.

Both constructors take the grid size and the wrap flag, indicating if your grid is flat or donut shaped. The second constructor brings a number of cells to life, within a bounding rectangle of start_width by start_height. You can make the first delegate to the second, to save duplicating code.

Alive/GameOfLife/GoL.cpp
```cpp
World::World(size_t max_x, size_t max_y, bool wrap) :
  World(max_x, max_y, wrap, max_x, max_y, 0)
{
}
```

In your second constructor, bring the requested number of cells to life. You want these to start inside your paper bag, so you need to check they fit in a start_width * start_height rectangle, and throw an exception if they don't. Bring the cells to life by filling the first few items with true. You can then shuffle the first start_width * start_height cells, to randomize your setup:

Alive/GameOfLife/GoL.cpp
```cpp
World::World(size_t max_x, size_t max_y,
             bool wrap,
             size_t start_width, size_t start_height,
             size_t number) :
  max_x(max_x),
  max_y(max_y),
  state(max_x*max_y, false),
  wrap(wrap)
{
  if (number > start_width*start_height)
    throw std::invalid_argument("Start rectangle too small");
  if (number)
  {
    std::fill_n(state.begin(), number, true);
    std::random_device rd;
    std::mt19937 gen(rd());
    std::shuffle(state.begin(),
                 state.begin() + start_width*start_height,
                 gen);
  }
}
```

Look back to the dice rolling example on page 110 if you need a reminder of how to use the random number generators in C++.

Use the second constructor to create a random new World. Once you've found some interesting patterns, you can use the first constructor and Spark cells to replicate those patterns. For example, you can set up a World with the three cells in a line blinking from the diagram on page 151.

To see this in action, you need to make an animation of your CA. You will indicate which cells are Alive and then Update your grid. A cell state depends on the number of live neighbors, so you need to find these. Look back to the rules on page 150 for a reminder. Cells have eight neighbors—unless they are at the edge of a flat grid, in which case they have fewer. When you wrap, you have a torus, so every cell has eight neighbors, as you saw on page 151.

To provide both options, create walkNeighbors and walkNeighborsWithWrapping functions:

```cpp
void walkNeighbors(size_t x, size_t y, size_t max_x, size_t max_y,
        std::function<void(size_t, size_t)> action)
{
  if(y>0)
  {
    if(x>0) action(x-1,y-1);
    action(x,y-1);
    if(x<max_x-1) action(x+1,y-1);
  }
  if(x>0) action(x-1,y);
  if(x<max_x-1) action(x+1,y);
  if(y<max_y-1)
  {
    if(x>0) action(x-1,y+1);
    action(x,y+1);
    if(x<max_x-1) action(x+1,y+1);
  }
}

void walkNeighborsWithWrapping(size_t x, size_t y,
        size_t max_x, size_t max_y,
        std::function<void(size_t, size_t)> action)
{
  size_t row = y>0? y-1 : max_y -1;
  action(x>0? x-1 : max_x - 1, row);
  action(              x, row);
  action(x<max_x-1? x + 1 : 0, row);
  row = y;
  action(x>0? x-1 : max_x - 1, row);
  action(x<max_x-1? x + 1 : 0, row);
  row = y<max_y-1? y+1 : 0;
  action(x>0? x-1 : max_x - 1, row);
  action(              x, row);
  action(x<max_x-1? x + 1 : 0, row);
}
```

Both functions take an action, so you have flexibility. You can then perform any action you like for each neighbor. You can count how many are alive for your Update, or count them in some tests. To count how many are alive, use this lambda:

```
size_t countAlive = 0;
walkNeighbors(x, y, max_x, max_y,
  [&](size_t xi, size_t yi)
  {
    countAlive += Alive(xi, yi);
  });
```

Now you can Update your World. You need to apply the rules to work out if a given cell will StayAlive. A cell with the right number of live neighbors will StayAlive or even come to life. You use your neighbor walking functions like this:

Alive/GameOfLife/GoL.cpp
```
void World::Update()
{
  std::vector<bool> new_state(max_x*max_y, false);
  for (size_t y = 0; y<max_y; ++y)
  {
    for (size_t x = 0; x<max_x; ++x)
    {
      new_state[y*max_x + x] = StayAlive(x, y);
    }
  }
  state.swap(new_state);
}

bool World::StayAlive(size_t x, size_t y) const
{
  size_t countAlive = 0;
  if (wrap)
    walkNeighborsWithWrapping(x, y, max_x, max_y,
      [&](size_t xi, size_t yi)
      {
        countAlive += Alive(xi, yi);
      }
    );
  else
    walkNeighbors(x, y, max_x, max_y,
      [&](size_t xi, size_t yi)
      {
        countAlive += Alive(xi, yi);
      });
  if (Alive(x, y))
  {
    return countAlive == 2 || countAlive == 3;
  }
```

```
    else
        return countAlive == 3;
}
```

To Update your CA make a new grid, to implement the canonical offline update. Decide if a cell should StayAlive and set the new_state appropriately. When you check the cell's new state in StayAlive, you count the live neighbors and take the appropriate action based on the current cell's state. If you tried an online method instead, the cells' states would change as you made your updates. By making a new_state grid, the cell states change together in a batch.

Now you're ready to make a world. It's important to remember that you must leave an edge above the top of the bag, or else nothing can get out. You also need at least three live cells next to each other to start new life, so pick a large enough starting number. Include the game of life header, and leave it running:

```
int main(int argc, char** argv)
{
    const bool wrap = true;
    const size_t bag_width = 50;
    const size_t bag_height = 40;
    const size_t edge = 10;

    const size_t world_x = bag_width;
    const size_t world_y = bag_height + edge;
    const size_t number = 800;

    World world(world_x, world_y, wrap, bag_width, bag_height, number);
    while(true)
        world.Update();
}
```

You can put this code inside main, but you won't see much. If you call Update in an SFML main loop you can display what's going on. Choose a shape and color for the live cells. A small cyan circle works well:

```
sf::CircleShape shape(5);
shape.setFillColor(sf::Color::Cyan);
```

Set up a project using the SFML library and make a window as you've done before, for example on page 111. Draw a bag as you did earlier on page 115 and Update your world in the main loop:

Alive/Alive/main.cpp
```
void draw(World & world, size_t edge)
{
    const float cell_size = 10.0f;
    const float width = world.Width() * cell_size;
    const float margin = edge * cell_size;
    const float line_width = 10.0f;
```

```
const float height = world.Height() * cell_size;
const float bag_height = (world.Height() - edge) * cell_size;
const auto bag_color = sf::Color(180, 120, 60);

sf::RenderWindow window(
  sf::VideoMode(
    static_cast<int>(width + 2 * margin),
    static_cast<int>(height + margin)),
  "Game of Life");

bool paused = false;
while (window.isOpen())
{
  sf::Event event;
  while (window.pollEvent(event))
  {
    if (event.type == sf::Event::Closed)
      window.close();
    if (event.type == sf::Event::KeyPressed)
      paused = !paused;
  }

  window.clear();
  drawBag(window,
      line_width,
      margin,
      bag_height,
      width,
      cell_size,
      bag_color);

  draw_world(world, cell_size, height, edge, window);

  window.display();
  std::this_thread::sleep_for(std::chrono::milliseconds(100));
  if(!paused)
  {
    world.Update();
  }
}
}
```

The draw_world function displays the live cells:

Alive/Alive/main.cpp
```
void draw_world(const World & world,
                float cell_size,
                float height,
                size_t edge,
                sf::RenderWindow & window)
{
  for (size_t y = 0; y<world.Height(); ++y)
```

```
    {
      for (size_t x = 0; x<world.Width(); ++x)
      {
        if (world.Alive(x, y))
        {
          sf::CircleShape shape(5);
          shape.setFillColor(sf::Color::Cyan);
          shape.setPosition((x + edge) * cell_size, height - y * cell_size);

          window.draw(shape);
        }
      }
    }
}
```

You're using conventional mathematical coordinates for your algorithm, so you need to subtract your y coordinate from the height to get the position on the window. Look back to the figure on page 62 for a reminder. By making a world and calling your draw function, you have a working Game of Life.

Did It Work?

The grid size and number of live starting cells affects your World. If the live cells are too sparse or over-crowded, they die off. Starting with 800 alive for a 40x50 grid was a good compromise.

You get different results each time if you randomly position your live cells. Did you get some stable patterns, oscillators, or spaceships? You can see six blinker oscillators in the following screenshot along with several stable patterns. The grid has settled to changing between the two states shown:

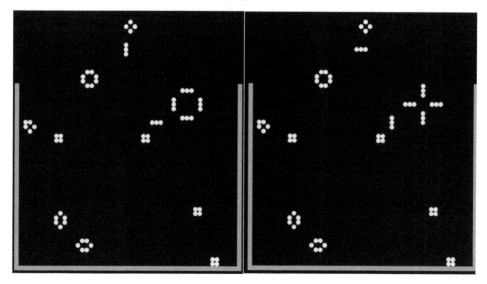

There are many known oscillators and spaceships. Seeing a few of them suggests your code is working. An online Wiki collates many patterns and facts if you want to explore further.[10] For example, no one has found an eight period spaceship yet. The *caterpillar* pattern contains eleven million cells. That won't fit in your small grid.

Your relatively small grid tends to stabilize quite quickly. If you allow the World to wrap around, it may change for a while longer. Whichever you choose, you are likely to end up with some live cells outside of your bag. You can also start with an empty World and set up a glider or other pattern. On a flat grid, it will get stuck in the bag if it glides down. If you wrap this into a torus, it will move forever. The starting state affects whether it glides up or down. If you get it the right way around, it will end up outside of your bag:

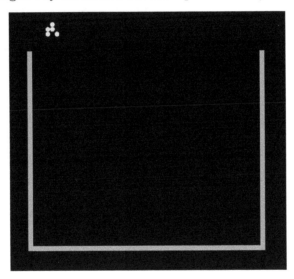

Over to You

Your bounded grid, whether flat or a torus, doesn't allow many patterns to emerge. To find more patterns, you really need an infinite grid. To achieve that, you need to change your data structure for tracking the cells' state. You can try that extension yourself. Alternatively, you can browse an online catalogue for known patterns and even run their code to try to discover new patterns.[11]

---

10. http://www.conwaylife.com/wiki/Main_Page

11. http://catagolue.appspot.com/

You saw Langton's ant earlier, and now you can try coding another CA. You can even make up your own rules and use a variety of colors for cells rather than only distinguishing between alive or dead:

- Choose a color based on the majority of the neighbors' colors.
- Choose a color based on the average of the neighbors' colors.
- Cycle through some colors, leaving a cell as it is or matching a neighbor if that color is next in your cycle.

Wireworld has four states represented by different colors.[12] Wireworlds are Turing complete and can generate logic gates. Once you can build logic gates, you can build a computer. In theory, you can even make a genetic algorithm build a Wireworld to perform specific tasks.

You can also see your rules as ways to make decisions. With several colors or states, you can build rules for a finite state machine. If you provide some form of feedback or reinforcement, you build a learning automata, which takes a step toward *reinforcement learning*.[13] Reinforcement learning plays a pivotal part in some recent trends in machine learning—for example, AlphaGo, the first computer program to beat a professional Go player.[14] John McCarthy, mentioned on page 1, noted how hard writing a program to win Go was, saying, "Sooner or later, AI research will overcome this scandalous weakness."[15]

In the next chapter, you'll find the best starting configuration for two more CAs using a genetic algorithm. Your previous genetic algorithm in Chapter 3, *Boom! Create a Genetic Algorithm*, on page 33 always split your chromosome in the middle, since you only had two bits of information. Next time, you will have more than two bits of information, so need a different crossover scheme. This will solidify your knowledge of genetic algorithms, taking you near the end of your machine learning exploration.

---

12. en.wikipedia.org/wiki/Wireworld
13. en.wikipedia.org/wiki/Learning_automata
14. https://deepmind.com/research/alphago/
15. http://jmc.stanford.edu/artificial-intelligence/what-is-ai/index.html

CHAPTER 9

Dream! Explore CA with GA

In the previous chapter, you built a cellular automaton and made cells come to life outside of a paper bag. Some of the patterns were stationary, some cycled through states, and some glided through space. You couldn't ensure you got specific patterns by starting with a random selection of live cells. Some emerged if you got lucky. You were also able to spark specific cells to life to see a known pattern, like the glider introduced on page 148 . But what if you want all of the cells above the paper bag come to life?

Imagine a rudimentary cellular automaton operating in one dimension. It takes a row of cells and returns that exact same row. If you stack up the rows, one at a time in a paper bag, you'll eventually have a row above the bag. Some might be alive. You can even work out how to get all of the cells alive above the bag—for example, you can start with all of the cells alive. Can you get your computer to learn how to do this? If it manages with this straightforward problem, can it find a good starting row for a more complicated CA? Let's see.

In this chapter, you'll start with a rudimentary rule and build a genetic algorithm to find the best starting configuration. You'll try your algorithm with *elementary cellular automata* too.[1] These also operate in one dimension and stack up rows. Your genetic algorithm will manage to turn on several cells for some of these, but not always. You can try to make it kill all of the cells off instead to see if it can manage that. You'll then create a random cellular automaton, choosing how to change a row on the fly. Your genetic algorithm will struggle to find a good starting point for this CA. Being aware of when things can go wrong is important. Machine learning is not a silver bullet capable of solving any problem. You always need to think about what you're trying to do.

---

1.   http://mathworld.wolfram.com/ElementaryCellularAutomaton.html.

Your first genetic algorithm in Chapter 3, *Boom! Create a Genetic Algorithm*, on page 33 had two parameters, angle and velocity, so you split solutions down the middle to breed new solutions. When your row is longer than two cells you have more options, so you will learn other crossover schemes. You'll also implement a tournament selection, since you have used roulette wheels twice now in *Using the Fitness Function*, on page 43 and *Your Mission: Lay Pheromones*, on page 80. Then, you'll have a solid grounding in genetic algorithms, and you'll see how the same overall approach applies to a completely different problem. Such a general-purpose machine learning algorithm is a *meta-heuristic*. By encoding your problem suitably and choosing a sensible fitness function you can guide your algorithm toward solving many different problems.

Your Mission: Find the Best

You'll start with the simple rule—any row stays as it is. You stack up copies of this row until you're above the paper bag:

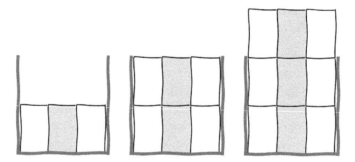

Will your genetic algorithm be clever enough to switch on all of the cells to begin with? This challenge is a variant of the *OneMax* problem.[2] You don't need an *evolutionary algorithm* (EA) such as a GA to solve this. However, it does illustrate the characteristics of an EA, and the fitness function is easy to understand.

You can express your problem in formal terms. Find a bit pattern

$$\overline{x} = (x_1, x_2, ..., x_n), x_i \in \{0, 1\}$$

that maximizes

$$\sum_{i=1}^{n} x_i$$

---

2.   http://tracer.lcc.uma.es/problems/onemax/onemax.html.

The formalism for such a trivial problem might seem over-the-top. You often see machine learning problems expressed in formal language, so it's worth being familiar with. In fact, using a genetic algorithm for this simple problem is over-engineered. You can solve this problem without a computer. However, learning how complicated things work with simple examples is useful when you're just starting out.

You'll implement the rudimentary CA as a function, which takes a row and returns that exact same row. Once you build a genetic algorithm to explore the OneMax problem, you can use the exact same GA for other functions.

The genetic algorithm starts with an initial population, creating rows of randomly selected 0s and 1s. As you know from your cannon GA on page 39 the algorithm runs for a few epochs, trying to improve:

```
items = 25
epochs = 20
generation = random_tries(items)
for i in range (1, epochs):
  generation = crossover(generation)
  mutate(generation)
display(generation)
```

Previously, the population had twelve items and ran for ten epochs. This time, you'll need more items, allowing your algorithm to try more starting rows, and more epochs, giving it more time to learn. The next section shows you how to implement crossover, which needs a fitness function, and mutation. You're trying to end up with a row of 1s for each of your CAs. Your fitness function can therefore count the 1s at the top. More is better.

You need to decide how long your rows are. In the previous chapter, you allowed this to vary. This time, keep that fixed to 32 cells, so you have fewer options. You're going to explore three different CAs, so you have plenty to try out already. The same method works on different sizes, so try them too. However, you then need to experiment with the population size and number of epochs.

With OneMax under your belt, you'll pick an *elementary cellular automaton* (ECA) and try to find the best way to start your cells. Your first rule left the cells as they were each time. This time, however, each cell might change state.

An ECA is one of 256 rules. The rule tells you how to change a cell by considering its state and that of its two neighbors. The edge cells only have one neighbor to consider, so you treat the non-existent neighbors as dead. There are eight possible states as shown in the table on page 166.

Left cell	This cell	Right cell
Alive	Alive	Alive
Alive	Alive	Dead
Alive	Dead	Alive
Alive	Dead	Dead
Dead	Alive	Alive
Dead	Alive	Dead
Dead	Dead	Alive
Dead	Dead	Dead

By selecting a 0 or 1 for each of these eight states, you generate an eight-bit number between 0 and 255. The number, in binary, shows you what to do with any input row. Let's try an example.

Consider the number 30. In binary, that number is 00011110:

$$0 \times 128 + 0 \times 64 + 0 \times 32 + 1 \times 16 + 1 \times 8 + 1 \times 4 + 1 \times 2 + 0 \times 1$$

The leftmost bit is 0, so the first state in the table—all cells on—maps to 0. This tells you that any cell that's on, whose neighbors are both on, turns off under rule 30. The next bit is also 0, so the second row in the table maps to off. And so on, like this:

$$111 \to 0, 110 \to 0, 101 \to 0, 100 \to 1, 011 \to 1, 010 \to 1, 001 \to 1, 000 \to 0$$

Or, as a picture:

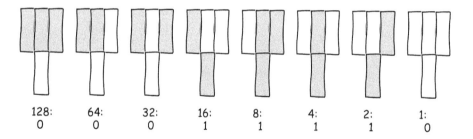

By convention, the ECA places rows of cells underneath one another. You'll start at the bottom and stack up your rows instead, so they end up outside of the paper bag. Your genetic algorithm will attempt to turn on as many cells as possible in the final row for any choice of the 256 rules. You'll make the target configurable, so you can try to get your GA to turn cells off instead. Some rules are more amenable to one target than the other.

You'll then try your meta-heuristic on a final CA. The ECA maps each cell in a row to a new state. An even more general CA takes a row and maps that to another row. This has considerably more than 256 possible mappings, for a row wider than 8 cells. Rather than enumerating these in advance, you can dynamically build a lookup table. If a row has an entry in your table, the corresponding value tells you which row comes next. If not, you generate a new output row, and add this to your table. You'll still stack up the rows and see what your GA manages.

Can it turn all of the cells on at the top? Will any randomly generated row even have all the cells turned on? Trying to find the best starting point is a challenge for your genetic algorithm. Each time you try a new row, you might generate a new mapping for your rule. Trying to find the best solution to a problem changing under your feet is difficult. Some machine learning algorithms can cope with dynamic environments. However, this variant is dynamic and completely random. You can't always find an answer to a problem with machine learning. Some problems don't have solutions, but exploring a problem to see what happens can be informative. This can give you a feel for complex problems, and might help you realize something is not possible.

How to Explore a CA

You'll try to find the best starting row for each CA using a genetic algorithm. You need a population of rows, each formed from random 0s and 1s, to get started. You run your automaton on each row, and see how many cells are alive when they get above the paper bag. Regardless of the CA you're exploring, you need to breed new solutions for your chosen number of epochs, mutating a few rows in your population as you go. Over time, the GA might find better starting rows.

To breed, you crossover the information from two parents and form a new potential solution. You want your algorithm to breed better solutions over time. For this problem, you'll use *tournament selection*, making three solutions picked at random compete. You can try other numbers, but three works. These three starting rows will battle it out. The tournament winner has the most cells alive at the top of the paper bag. The first tournament gives you Mum. Rinse and repeat to get a Dad. You can then breed a new starting row, combining the parents' information.

Previously, you had two variables, angle and velocity, so you took one from Mum and one from Dad, on page 46. Now you have several variables—one per cell. They are bits rather than real values, however, you have more ways to split solutions for crossover now. The simplest scheme splits in the middle again, or close to the middle for an odd row length, joining half from each, as shown in the next figure:

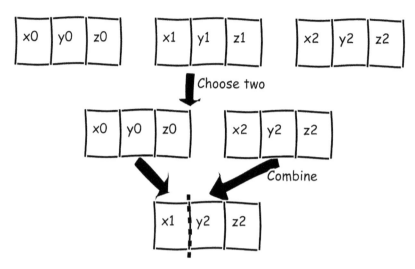

Your crossover operation might now make worse children. Look at what happens when you split the following parents down the middle and take half from each:

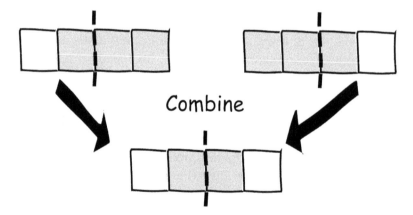

Mum and Dad both have three live cells, but the child only has two. You were hoping to get better solutions as your algorithm ran. The algorithm will weed out worse children next time a tournament happens, but some implementations keep an eye out for situations like this. You'll run a tournament between Mum, Dad, and child to decide which one joins the next generation. Since you're after better solutions, you'll also keep the best row from the previous generation each time. You learned about this elitist selection on page 45 when you first made a GA, but didn't use it back then. Try it out now.

So, where will you split the rows? You can split at a single point, either a pre-chosen place, like the middle, or by picking a random point each time. You can even split the row in several places and splice together a new solution. More complicated approaches tend to create a greater variety of solutions in a population. Try the simplest approach first, splitting your row in one place. This will either be in half or at a random point, driven by parameter, allowing you to experiment. Feel free to try other extensions when you have this working.

To complete the algorithm you need to mutate the rows. A simple approach is probabilistically picking a cell and changing its state. You can extend this to mutate more than one cell. You need to decide a mutation rate. Keeping the best row in each generation and only selecting better children narrows down the solutions your algorithms find. This is known as premature convergence, so you need to be careful with parameters for larger problems. You therefore need quite a high mutation to give it a helping hand—try 50% initially. This will encourage some variety in the population. Do try turning it off completely, or always mutating too, just to see what happens.

You now have a way to start, and you know how to implement crossover and mutation as your algorithms learn. Time to code it.

Let's Find the Best Starting Row

Each of your CAs will use a Row of 32 cells:

```
typedef std::array<bool, 32> Row;
```

The array has a fixed size. You can use a vector, as you did in the previous chapter, if you want to dynamically change the row size. You'll need to experiment with the parameters for your GA if you change this.

Each CA works via a rule acting on a Row. If you make an abstract Rule, you'll be able to polymorphically swap your CA for use in the same genetic algorithm. To do this, use a virtual operator() to transform a Row:

```
Dream/GACA/rule.h
class Rule
{
public:
    virtual Row operator()(const Row & cells) const = 0;
};
```

To make a concrete rule, you override the virtual function. Your first CA will use a StaticRule:

```
Dream/GACA/rule.h
class StaticRule : public Rule
{
public:
  virtual Row operator()(const Row & cells) const
  {
    return cells;
  }
};
```

As you can see, this keeps a row exactly as it is. To make this rule, use a shared_ptr, so you can select the type of Rule at runtime when you run your GA:

```
std::shared_ptr<Rule> rule = std::make_shared<FullRule>();
```

Let's build the parts you need for the GA. As you did before in *Let's Make Cellular Automata*, on page 154, make a World class to try out your problem:

```
Dream/GACA/GACA.h
class World
{
public:
  World(const Rule & rule, Row row, size_t height);
  void Reset(Row row);

  bool Alive(size_t x, size_t y) const;
  size_t Height() const {  return height; }
  size_t Width() const { return row.size(); }
  Row State() const { return row; }
  Row StartState() const { return history[0]; }
private:
  const Rule & rule;
  Row row;
  size_t height;
  std::vector<Row> history;
  void Run();
};
```

Your GA will make a few of these, trying to improve over time. You want to know how many cells in the top row are alive or dead, so Run the rule until your rows reach the chosen height, stacking up a history as you go:

```
Dream/GACA/GACA.cpp
void World::Run()
{
  while (history.size() < height)
  {
    history.push_back(row);
    row = rule(row);
  }
}
```

You'll make worlds compete in a tournament. If you run all of the updates in the constructor, you have the top row's State to hand for the tournaments:

Dream/GACA/GACA.cpp
```
World::World(const Rule & rule, Row row, size_t height) :
  rule(rule),
  row(row),
  height(height)
{
  Run();
}
```

As your GA runs, it will make new worlds and mutate these from time to time by changing the starting row via a reset method:

Dream/GACA/GACA.cpp
```
void World::Reset(Row new_starting_row)
{
  row = new_starting_row;
  history.clear();
  Run();
}
```

The genetic algorithm needs a Population of worlds to breed and mutate. You can use a vector for this:

```
typedef std::vector<World> Population;
```

The GA works on an initial random population, so you need to generate a Row for each World randomly.

You can generate rows like this:

Dream/GACA/cells.h
```
class RowGenerator
{
public:
  RowGenerator(std::random_device::result_type seed) :
    gen(seed),
    uniform_dist(0, 1)
  {
  }
  Row generate();
private:
  std::default_random_engine gen;
  std::uniform_int_distribution<size_t> uniform_dist;
};
```

The uniform_dist generates a 0 or 1. You use the number to determine the state of each cell in a Row:

```
Dream/GACA/cells.cpp
Row GACA::RowGenerator::generate()
{
  Row cells;
  for (size_t i = 0; i < cells.size(); ++i)
  {
    cells[i] = (uniform_dist(gen) == 1);
  }
  return cells;
}
```

Select the rule to use, either the StaticRule or the ECA or dynamic rules you'll see shortly. Decide the population size and how many updates to get to the top of the paper bag to create a start Population for your GA:

```
Dream/Dream/main.cpp
Population start(const GACA::Rule & rule,
  size_t size,
  size_t updates)
{
  std::random_device rd;
  Population population;
  RowGenerator cell_generator(rd());
  for (size_t i = 0; i<size; ++i)
    population.emplace_back(rule, cell_generator.generate(), updates);
  return population;
}
```

Your GA will now try to improve on your initial population. You need to breed new starting rows and mutate them from time to time. You breed a row using information from two parents:

```
Dream/GACA/GACA.cpp
Row GACA::breed(Row mum, Row dad, size_t split)
{
  Row new_row;
  auto it = std::copy(mum.begin(), mum.begin() + split, new_row.begin());
  std::copy(dad.begin() + split, dad.end(), it);
  return new_row;
}
```

The split point might be in the middle or anywhere along a row. Making this configurable allows experimentation. In order to select the parents, you need a crossover implementation.

Crossover

Crossover needs a constructor and an operator to generate the next Population using a tournament function to select parents:

```
class Crossover
{
public:
    Crossover(std::random_device::result_type seed,
      size_t population_size,
      const Rule & rule,
      size_t updates,
      bool middle,
      bool target);
    Population operator()(const Population & population);
        const World & Crossover::tournament(const World & world1,
                const World & world2,
                const World & world3) const;
private:
    std::default_random_engine gen;
    std::uniform_int_distribution<size_t> uniform_dist;
    std::uniform_int_distribution<size_t> split_dist;
    const Rule & rule;
    const size_t updates;
    const bool middle;
    const bool target;
};
```

Let's work through the details. First, make a constructor:

```
Crossover::Crossover(std::random_device::result_type seed,
    size_t population_size,
    const Rule & rule,
    size_t updates,
    bool middle,
    bool target) :
  gen(seed),
  uniform_dist(0, population_size - 1),
  split_dist(0, std::tuple_size<Row>::value - 1),
  rule(rule),
  updates(updates),
  middle(middle),
  target(target)
{
}
```

The constructor sets up two random uniform integer distributions. You'll use the uniform_dist to choose possible parents, so you need it to generate an index into your population. Any number from 0 to one less than the population_size works. These will compete in a tournament. You'll run this twice to get two parents who breed. You made the split point used in breeding configurable, so store your choice of middle or not, to use later. To vary the point, you use split_dist to generate a number between 0 and one less than the size of the Row, giving you an index into the Row. You also choose a target, so you can aim for everything on or off in the top row.

The Crossover operator makes a new Population using the current one:

```
Dream/GACA/GACA.cpp
Population Crossover::operator()(const Population & population)
{
  const size_t size = population.size();
  if (size-1 != uniform_dist.max())
  {
    std::stringstream ss;
    ss << "Expecting population size " << uniform_dist.max()
        << " got " << size;
    throw std::invalid_argument(ss.str());
  }
  Population new_population;
  auto best_world = best(population, target);
  new_population.push_back(best_world);

  while(new_population.size() < size)
  {
    const World & mum = tournament(population[uniform_dist(gen)],
                          population[uniform_dist(gen)],
                          population[uniform_dist(gen)]);
    const World & dad = tournament(population[uniform_dist(gen)],
                          population[uniform_dist(gen)],
                          population[uniform_dist(gen)]);

    Row new_row = breed(mum.StartState(), dad.StartState(),
                    middle ?
                        std::tuple_size<Row>::value / 2
                    : split_dist(gen));
    World child(rule, new_row, updates);
    World winning_world(rule,
            tournament(child, mum, dad).StartState(),
            updates);
    new_population.push_back(winning_world);
  }
  return new_population;
}
```

You're keeping the best World each time on line 12, so you need to find it:

```
Dream/GACA/GACA.cpp
const World& GACA::best(const Population & population, bool target)
{
  return *std::max_element(population.cbegin(), population.cend(),
    [&](const World & lhs, const World & rhs)
    {
      return fitness(lhs.State(), target)
                   < fitness(rhs.State(), target);
    });
}
```

By ordering items from smallest to largest, max_element can find the element with the maximum fitness.

Your new_population now contains your best World, but it needs more items. You therefore breed new items until you have the required size on line 15 in the crossover operator. The parents come from your tournament of three worlds picked at random. You compare the fitness of each world's final State:

```
Dream/GACA/GACA.cpp
const World & Crossover::tournament(const World & world1,
    const World & world2,
        const World & world3) const
{
  size_t alive1 = fitness(world1.State(), target);
  size_t alive2 = fitness(world2.State(), target);
  size_t alive3 = fitness(world3.State(), target);
  if(alive1 < alive2)
  {
    if(alive1 < alive3)
      return alive2 < alive3 ? world3 : world2;
    return world2;
  }
  if(alive2 < alive3)
    return alive1 < alive3 ? world3 : world1;
  return world1;
}
```

You don't have to stick with three competitors, though this is common. Code alternatives if you want. You run a tournament twice to pick a mum and dad. The winner has the best fitness. Since you want to maximize the cells in the final Row with the desired target, you can count how many cells match this target:

```
Dream/GACA/GACA.cpp
size_t GACA::fitness(const Row & row, Row::value_type target)
{
  return std::count(row.begin(), row.end(), target);
}
```

You saw how a child might have worse fitness than the parents on page 168. You therefore run a tournament between the parents and child to decide which enters the new_population on line 30 in the crossover operator on page 174.

You can now breed as many items as you need. By rejecting worse children and keeping the best row each time, you have narrowed down your population. However, you can add a bit of variety with Mutation.

Mutation

Make a class for Mutation:

Dream/GACA/GACA.h
```
class Mutation
{
public:
    Mutation(std::random_device::result_type seed, double rate);
    Row operator()(Row cell);
private:
    std::default_random_engine gen;
    std::uniform_int_distribution<size_t> uniform_dist;
    const double rate;
};
```

You can get away with a single uniform_int_distribution to control whether a row is mutated, and which cell to mutate, by picking a number from 0 to the cell size minus 1. For a rate of 50%, (0.5), any number smaller than half the row size means mutate. For a rate of 25%, any number smaller than a quarter of the row size means mutate. If you decide to mutate the Row, pick a cell using the same random number generator, and toggle the value:

Dream/GACA/GACA.cpp
```
Row Mutation::operator()(Row row)
{
  auto maybe = uniform_dist(gen);
  if (maybe < rate*row.size())
  {
    auto index = uniform_dist(gen);
    row[index] = !row[index];
  }
  return row;
}
```

The GA Itself

To run your whole genetic algorithm, use the simple rule you saw earlier on page 170, and decide the parameters you'll use:

```
Dream/Dream/main.cpp
GACA::World ga_ca(const GACA::Rule & rule,
  size_t size,
  double rate,
  size_t epochs,
  size_t updates,
  bool middle,
  bool target)
{
  Population population = start(rule, size, updates);

  std::random_device rd;
  Mutation mutation(rd(), rate);
  Crossover crossover(rd(), size, rule, updates, middle, target);
  for(size_t epoch = 0; epoch < epochs; ++epoch)
  {
    population = crossover(population);
    for(auto & world : population)
      world.Reset(mutation(world.StartState()));

    const World & curr_best_world = best(population, target);

    auto alive = fitness(curr_best_world.State(), target);
    std::cout << epoch << " : " << alive << '\n';
  }
  const World & best_world = best(population, target);
  std::cout << "Final best fitness "
            << fitness(best_world.State(), target) << '\n';
  return best_world;
}
```

In each epoch, you obtain a new population from crossover. You then use the reset method to pick up changes from mutation. You can report the fitness of the best world each time to see if your algorithm is learning.

Once you've found the best World, you can draw what happens using the SFML you first saw on page 110. Draw a bag as you did before on page 115 and represent live cells using a small, cyan sf::CircleShape:

```
Dream/Dream/main.cpp
Line 1  void draw(World & world)
    -   {
    -     const size_t edge = 15;
    -     const float cell_size = 10.0f;
    5     const float width = world.Width() * 2*cell_size;
    -     const float margin = edge * cell_size;
    -     const float line_width = 10.0f;
    -     const float height = (world.Height() + edge)* cell_size;
    -     const float bag_height = world.Height() * cell_size;
    10    const auto bagColor = sf::Color(180, 120, 60);

    -     const int window_x = static_cast<int>(width + 2* margin);
```

```
     const int window_y = static_cast<int>(height + margin);
     sf::RenderWindow window(sf::VideoMode(window_x, window_y), "Dream!");

     bool paused = false;
     size_t row = 1;
     while (window.isOpen())
     {
       sf::Event event;
       while (window.pollEvent(event))
       {
         if (event.type == sf::Event::Closed)
             window.close();
         if (event.type == sf::Event::KeyPressed)
           paused = !paused;
       }

       window.clear();
       drawBag(window,
         line_width,
         margin,
         bag_height,
         width,
         cell_size,
         bagColor);
       for(size_t y=0; y<row; ++y)
       {
         for(size_t x=0; x<world.Width(); ++x)
         {
           if(world.Alive(x, y))
           {
             sf::CircleShape shape(5);
             shape.setFillColor(sf::Color::Cyan);
             shape.setPosition(x * 2*cell_size + margin, height - y * cell_size);
             window.draw(shape);
           }
         }
       }

       window.display();
       std::this_thread::sleep_for(std::chrono::milliseconds(100));
       if(!paused && (row < (world.Height() + edge/2.0)))
         ++row;
     }
   }
```

You can stack up the rows one at a time, to show how they change. Start with the first row on line 17. As you draw each update, go up to the current row on line 37. In order to see your rows stack up, increment this in the loop, on line 54. Keep going for a few extra rows above the paper bag, to half the edge you left at the top, to make clear what your GA has achieved.

If you draw cells right next to one another, they look a bit squashed. This code, therefore, leaves a small gap between each cell, on line 45, by setting the x position to double the cell_size.

You can draw any World using this function. See how your GA does with the OneMax problem, and then build your next CA.

Elementary Cellular Automata

An elementary cellular automaton has a rule number. Derive a new class from the abstract Rule, saving the rule number:

Dream/GACA/rule.h
```
class ECARule : public Rule
{
public:
  explicit ECARule(size_t rule) : rule(rule)
  {
  }
  virtual Row operator()(const Row & cells) const
  {
    Row next;
    next.fill(false);
    for (size_t i = 0; i<std::tuple_size<Row>::value; ++i)
    {
      std::bitset<3> state = 0;
      if (i>0)
          state[2] = cells[i - 1];
      state[1] = cells[i];
      if (i<std::tuple_size<Row>::value - 1)
          state[0] = cells[i + 1];

      next[i] = rule[state.to_ulong()];
    }
    return next;
  }
private:
  const std::bitset<8> rule;
};
```

To find the next row from the current row of cells, you walk through each cell in the row and decide its state. You make a three bit number (state) from the current cell and its neighbors. The value of this bit in the rule tells you if a cell should live or die. Glance back at the picture on page 166 for a reminder of the mapping between the bits and the next cell state. Once you have set all of the bits in the next row, you return it.

Display what happens for a given rule when you start with the middle cell on:

```
Dream/Dream/main.cpp
void eca_display(size_t number, double rate, size_t height)
{
  using namespace GACA;
  std::shared_ptr<Rule> rule = std::make_shared<ECARule>(number);
  Row cell;
  cell.fill(false);
  cell[cell.size()/2]=true;
  World world(*rule, cell, height);
  draw(world);
}
```

You can try your GA on a random rule by selecting an integer from 0 to 255:

```
std::random_device rd;
std::default_random_engine gen(rd());
std::uniform_int_distribution<size_t> uniform_dist(0, 255);
size_t number = uniform_dist(gen);
auto rule = std::make_shared<ECARule>(number);
```

Send this rule into ga_ca on page 177 and see what happens.

Dream Up a Rule

For your third CA, you need to build up a lookup table of randomly generated rows, mapping an input row to one dreamt up on the spot if you've not encountered it before. To do this, you want something that generates a row at random, so you can repurpose the RowGenerator you saw on page 171. Make another concrete Rule:

```
Dream/GACA/rule.h
class DreamRule : public Rule
{
public:
  explicit DreamRule(std::random_device::result_type seed) :
      gen(seed)
  {
  }
  virtual Row operator()(const Row & cells) const;
private:
  mutable std::map<Row, Row> lookup;
  mutable RowGenerator gen;
};
```

Check if you've seen the Row before in your operator, and generate a new Row if not:

```
Dream/GACA/rule.cpp
Row DreamRule::operator()(const Row & cells) const
{
  auto it = lookup.find(cells);
  if (it != lookup.end())
    return it->second;
  Row return_cell = gen.generate();
  lookup[cells] = return_cell;
  return return_cell;
}
```

Creating one of these rules for your World is very simple:

```
std::random_device rd;
auto rule = std::make_shared<DreamRule>(rd());
```

You can try your GA on this rule now by using the ga_ca function on page 177.

You now have three types of CA implemented as rules:

- A rudimentary CA leaving a row as it is.

- An elementary CA, composed of an 8-bit integer mapping the current state of a cell and its neighbors to a new state.

- A dynamic CA, randomly choosing a new state for a row it hasn't seen .

Your genetic algorithm tries to have as many cells with your chosen target by the top of the paper bag. This does not always succeed. Let's investigate what happens in more detail.

Did It Work?

If you start with the simplest possible rule, and only have one item in the population, not much changes. Your tournament only has one competitor. The mutation gives it a chance to improve, but it might not manage to turn on all of the cells. Here's a typical run—it started with 12 live cells, and creeps up to 17 over time:

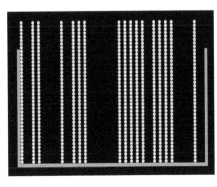

You need a larger population to allow your GA to try more rows. Experimentation suggested about 25 items works well. The algorithm can then manage to turn on all 32 bits within 20 epochs, like this:

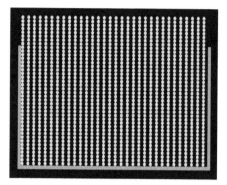

Your choice of whether to split in the middle during Crossover affects how often it succeeds. If you stick with the mid-point, your genetic algorithm can manage to turn on all of the bits, but sometimes only gets close. Choosing a random split point instead helps it do better more often. You saw how crossover can make things worse on page 168. Mutation, and running a tournament between a child and the parents can compensate for this, but using random split points helps even more.

What about the mutation rate? Turning the mutation off completely also tends to miss the best possible starting setup. The solutions improve over time, but you need the extra exploration, with a larger population and more epochs. If you set the rate to 100% (1.0) you might find the best possible setup, but mutate it and get something worse in the end. Using 50% works well.

Elementary Cellular Automata

You tried your ECA rules on rows with only the middle cell alive, on page 180. For rule 122, you build up a triangle pattern:

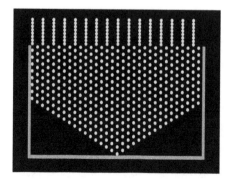

122 is 01111010 in binary. The initial row has most cells off, and the rule says they remain off, since the last bit is 0. One cell is on, with both neighbors off, giving state 010. This is the sixth row from the table on page 166, and the sixth bit in the rule is 0:

`111 : 0, 110 : 1, 101 : 1, 100 : 1, 011 : 1, 010 : 0, 001 : 1, 000 : 0`

Your starting cell, therefore, turns off. The left neighboring cell is in state 001, and the right cell is in state 100, so both turn on. You now have two cells in the same state as your original cell. This continues as you keep applying the rule giving the triangle. Looks like it worked.

You tried to find the best starting row with your genetic algorithm for a rule chosen at random from the 256 possible. Some rules tend to kill everything off, while others keep things the same. Your genetic algorithm may, therefore, not be able to find a row that ends up with every cell turned on. You can change the target value for the fitness to false so your algorithm tries to kill off all the cells instead. This can help for some rules—for example, rule 204.

If you aim to turn cells off for rule 204, the GA manages to empty the paper bag completely, with all of the cells turned off and remaining off. Not a major challenge for your algorithm—if it starts with one cell on in the middle, it remains in that state:

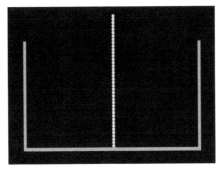

All your algorithm needs to do is turn off that cell, along with a few others, to find the best.

Dream Rules

Your algorithm has a hard time with the dynamic rules. A sample run is shown in the figure on page 184.

The rows are stacking up with no obvious pattern. Some cells are alive at the top, but not all of them. Since new rules are being made up as this runs, and they are entirely random, your algorithm is trying to make sense of *noise*. If

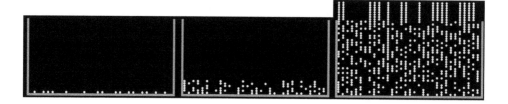

you debug, you'll see it frequently making new rules as your genetic algorithm tries to improve.

Trying an algorithm on random input data is common. If researchers claim that they can model data or solve a problem with greater accuracy than anyone else, they will sometimes use random data as an input to show that they have not made a mistake. You can work out how likely a good model will be on random data, so you can verify what should happen. In contrast, if an algorithm does not seem to be working at all, you can check if you get similar behavior for completely random inputs. If that happens, it can help you narrow down where real inputs are being transformed into nonsense.[3] Search for "Predictive Models on Random Data" to see several examples claiming machine learning can model random data. No algorithm can really model complete randomness. You need to assess if what you have done really works or even makes sense.

If you make your paper bag much shorter, there are far fewer possibilities, though you are still modeling noise. When the paper bag is two units high, your lookup table will map each starting row it tries to one new row. You might end up with a mapping to the row with all of the cells on, but you have no guarantee this will happen. Trying to turn all of the cells on might be impossible. If you make the bag height two, you're hoping for a rule mapping your starting cells to a row with all the cells on. If the bag is taller than two rows, you then hope to get a chain of rows ending up with all of the cells on. Each extra row multiplies up the number of possible outcomes exponentially. For a bag of two rows only, getting a row of cells on becomes less likely as the row gets wider as shown in the table on page 185.

You have less than a one in four billion chance of that happening for 32 cells across a row when the bag is two units high. Make it taller, and you have even less chance. There isn't much for your algorithm to learn.

---

3.   blog.slavv.com/37-reasons-why-your-neural-network-is-not-working-4020854bd607

Row Width	Possible rows	Probability all cells on
1	2: 0, 1	0.5
2	4: 00, 01, 10, 11	0.25
3	8: 000, 001, 010, 011, 100, 101, 110, 111	0.125
4	16: 0000,...	0.0625
5	32: 00000,...	0.03125
...	...	...
32	4294967296: ...	0.0000000002

Over to You

You have seen alternative ways to implement the crossover in genetic algorithms and used tournament selection. You also learned about the OneMax problem and elementary cellular automata. You even designed your own CA. You managed to find the best starting cells for the simplest problem, and explored the other setups. As an extension, you can try to find good rules for a fixed starting row. For the ECA you can look things up. It will take a while, but it is possible. Rather than the brute-force approach, you can try a genetic algorithm for this problem too.

You have also seen how machine learning algorithms sometimes fail to solve problems. Trying to model or learn how to improve when the problem is randomly changing is foolish. It isn't always clear in advance this is happening. Some people claim the stock market is just a random number generator so attempts to build trading strategies are simply modeling noise. Others disagree and do make money. Even if a problem seems very difficult, trying a random heuristic search can help you get a feel for what's going on.

If you want to try out a framework, the Distributed Evolutionary Algorithms in Python (DEAP) library has a walk-through of the OneMax problem.[4] You haven't used any framework in this book—you've been learning how the algorithms work with various challenges. Many frameworks exist. For real-world problems, you can use one instead of hand rolling your own implementation. Some, like DEAP, allow parallelization, which can speed up searches for complicated problems. You should be confident enough by now to try out existing frameworks or implement other algorithms.

You've seen a variety of different machine learning algorithms. Next, you'll wrap things up by looking at optimizations that don't use swarms, giving you

---

4. deap.readthedocs.io/en/master/examples/ga_onemax.html

a solid grounding in a wide gambit of machine learning approaches. You found the best starting cells in this chapter in order to escape a paper bag using a genetic algorithm. In Chapter 1, you hacked your way out of a paper bag without using any machine learning algorithms. You managed to find where the bag was with a decision tree in the next chapter. Since then, you've been trying out machine learning algorithms, escaping your paper bag over and over again. So, you might be wondering, how did you end up in a paper bag to begin with?

The next chapter will explore two ways to get to the bottom of the bag. First, you'll learn the hill climbing method and see how this gets stuck if you crumple up the paper bag. Then, you'll try *simulated annealing*, mimicking heat-treating metals, in effect allowing you to hit the paper bag with a virtual hammer until things work. These optimization methods can be adapted for use in training neural networks. If you wish to continue learning about these, you need to pick up a book or find a tutorial or framework. Now, let's round off your adventure in machine learning by getting into that paper bag.

CHAPTER 10

Optimize! Find the Best

In the previous chapter, you found the best starting configurations for cellular automata using genetic algorithms. You used a fitness function to assess how good a cell pattern was, and you programmed your way out of several paper bags. For this final exercise, you'll make your way into the paper bag instead.

Imagine you have a turtle who likes walking along the inside of a paper bag. This turtle is a little shy and is happiest when curled up in the bottom of the bag. Initially, this bag will be two-dimensional and almost rectangular. Later, however, you'll scrunch it up so you can learn how to find the best of several better hiding places without getting stuck in one. Although you'll see how to try more than two dimensions, you'll need to learn more mathematics to implement this extension on your own. In the meantime, you'll use the Python turtle package and two dimensions.

For this chapter, the plan is to get a turtle to walk along the bag's edge until he can't make it any farther down. You can start him inside of the bag, and get him to walk up—in this direction he's *hill climbing*. However, to get him into the bag, you need to find a minimum instead. You're still hill climbing, mathematically, even if your turtle is going downward. The direction is immaterial. Your goal is to *optimize* a function, finding a minimum within a range. You can use the same method to find a maximum too. In either case, you describe the bag's edges as a curve or function. The turtle steps left or right depending on which takes him farther down or up, depending on your directional goal. Regardless of the direction, the turtle will walk along the edge and stop when he can't go any farther. With an un-crumpled bag, this will succeed. With a slightly crumpled bag, however, he may miss a better spot since he can get stuck in a dip.

As you work through this problem, you'll see the difference between local minima and global minima; and like any good engineer, you'll find hitting things with a hammer can help. You'll use *simulated annealing* to bounce your turtle around. In metalwork, annealing heats, then cools a material to change physical properties, making metal less brittle or more ductile. As you hit the metal with a hammer, it becomes harder and might crack or break. Annealing avoids this problem by heating the metal. As it cools, there's less free energy, which means less movement of molecules. This changes the crystal structure of the material, making it more malleable. Simulated annealing borrows from the physics of the cooling process. To solve a problem you try alternatives, but unlike hill climbing, you accept a worse solution from time to time. As the systems cool, this becomes less likely. Changing a variable that represents temperature simulates annealing. You can then bounce the turtle around—as if you're hitting the world with a hammer–without breaking things, especially your turtle. In the long run, persuading the turtle to jump up a little bit can get him even lower into the bag. When that happens, you'll have a happy turtle.

Your Mission: Move Turtles

You used Python's Turtle package in Chapter 1 (on page 6), and you'll use it again now. You'll start by drawing a paper bag. All of the paper bags thus far have been rectangular. But now, you'll get to experiment with other shapes too.

You're going to put the turtle on the edge of the bag and let him walk. When he hill climbs or descends, he'll consider a step left or right. He chooses whichever goes down from his current position, picking the best spot. When he can't go down anymore, he'll stop.

This is another greedy algorithm. You already know greedy algorithms can get stuck or miss optimal solutions. You considered this on page 24 when picking the best feature to partition data in a decision tree. You'll see how easily the turtle gets stuck when he's greedy in this recipe too.

You also need to decide the turtle's step size. You can try a constant value initially, but you'll discover this sometimes gets him stuck as well. Of course, he doesn't always get stuck, though. With some paper bag shapes, your hill climbing turtle will find his way deep inside. Simple methods do sometimes work.

To avoid getting stuck, however, your turtle needs to be less greedy or try different step sizes. You can let him pick better spots, but shake things up once in a while by teleporting him to another place on the bag's edge. You have seen several ways to try something random to solve a problem. If you try a random step size and probabilistically pick a random spot when he gets stuck, you encourage him to explore more. He can then find a global minimum, rather than ending up in a local minimum.

Over time, you must decrease the step size and the chance of bouncing off to a different spot. Otherwise, the turtle may never settle down. He'll get worn out. To help with this problem, you can keep track of the temperature, which drops over time, and then use this to make jumps elsewhere less likely. Because if the turtle jumps, he might go up rather than down. A hill climbing turtle would never do this. The simulated annealing turtle will discover this helps to avoid local minima. By decreasing the step size, you let the turtle zone in precisely on the best spot.

You can try simulated annealing on several paper bags. You'll see one that doesn't have a single best spot, so a single lone turtle can't settle down in all of the equally good places. So, you can try a few turtles. With a bit of experimentation, you might get each of them to go to a different place.

How to Get a Turtle into a Paper Bag

You now have an overview of hill climbing and simulated annealing. Next, you'll see how to hill climb or descend a few different paper bags. Be forewarned, though, your turtles will get stuck in some of these. After that, you'll hit the setup with a hammer to simulate annealing. From there, you'll see how to use a temperature to model the energy of your system. This means you can probabilistically move the turtle to different spots. Ready for some hill climbing?

Climb Down a Hill

Your first recipe starts the turtle on the top-left edge of the bag. You'll move him a step, and then stop when the next step would be up:

```
pos = bag.left()
height = pos.y
while True:
  pos.step()
  if pos.y > height:
    break
  height = pos.y
```

Your turtle will start on the left and walk right, stopping when he's as far down as possible. You'll consider a step left or right when you code this. You'll also need to choose a step size. The turtle will move something like shown in the figure on page 190.

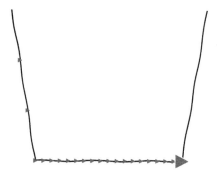

You've made things slightly easier by making the sides of the bag slant. This forms a proper mathematical function so one x value corresponds to exactly one y value. A rectangle has several y values on the right and left sides for the same x value, making it hard to step right or left. The turtle walks down the left side and along the bottom and then stops. Success. This works for paper bags with one low point or even a flat base. But what happens if the bag is slightly crumpled? Consider the paper bag in the next picture:

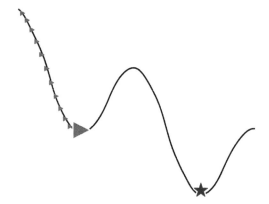

The paper bag is scrunched up to a curve:

$$y = 5 \times \cos(x) - x$$

The turtle, shown as triangles, walks down to a *local mimina*, shown by the larger triangle. At this point, he can't get lower with a small step left or right, so he stops where he is. Little does he realize, there's another minima just to the right—if only he went a little farther. In fact, there are lower points off the page too, but you can get the turtle to stay between the edges.

All the way back in Chapter 3 when we first looked at genetic algorithms, you saw how two solutions could be equally valid (as discussed on page 53). Your cannonballs could go right or left to get out of the paper bag. These were equally good directions. If this curve didn't tilt down but used

$$y = 5 \times \cos(x)$$

you have two equally good spots on the screen. You can adapt your algorithms, using two or more turtles, to find several minima. First, it's time to see how to use simulated annealing to get a single turtle farther down inside the paper bag so he finds a single minima.

Hit It with a Hammer

Instead of only picking a lower spot, you'll allow the turtle to jump to a completely different spot and continue his walk. Initially, this can happen frequently. Over time, he'll be less likely to jump. As the temperature cools in real annealing, the total amount of energy in the system decreases, so jumps of molecules get less likely. In simulated annealing, you also have a temperature cooling off, making the turtle jumps less likely.

Pick a temperature and subtract a small amount over time. Starting at 10.0 and decreasing by 0.1 works. This'll cool your system down slowly. Try other numbers too. This is an unusual cooling schedule but works for this problem. More common schemes use geometric reduction (multiplying) rather than additive methods. You often change the step size while the system cools, rather than at the end. You did these separately to learn one thing at a time. For this problem, nice and simple works. For more complicated problems you may need to try something involving a detailed mathematical analysis to see what works.

The temperature gives a probability of accepting a worse value. To be a probability you want a number between 0 and 1. If the temperature is zero or less, you won't move to a worse place. If the temperature is greater than zero, you need to decide how much worse the new place is. This will give you a negative number. You can then use the exponential function, to get a number between 0 and 1 as shown in the figure on page 192.

To measure how much worse positions are, you assign each an *energy*. This has a similar role to fitness functions or *loss* or *cost* functions in machine learning. A fitness function is bigger for better values. The loss, cost, and in this case, energy functions are bigger for worse values. For this problem, you can use the y coordinate as you have done many times before. If you subtract this from the turtle's current y coordinate, going uphill gives you a negative value. This makes higher values worse since you want the turtle to go down. If you want the turtle to go up, subtract the y coordinate from the paper bag height instead.

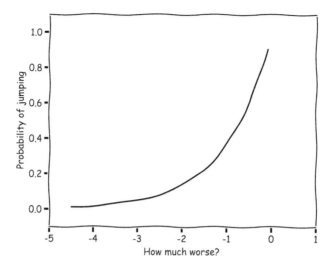

The turtle may choose a worse step. Or he may not. Over time, the temperature cools, making the transition to a worse state less likely. To decide whether to transition to this worse state, calculate

$$\text{probability} = e^{\left(\dfrac{\text{energy }(current)-\text{energy }(worse)}{\text{temperature}}\right)}$$

to obtain a number between 0 and 1. Dividing by the temperature gives you smaller probabilities over time. A point one unit higher has a difference in energy of -1. If the temperature is 0.5, you have

$$\text{probability} = e^{\left(\dfrac{-1}{0.5}\right)} = 0.135...$$

so you have a 13.5% probability of choosing this worse point. If the temperature drops to 0.25, this drops to under 2% since

$$\text{probability} = e^{\left(\dfrac{-1}{0.25}\right)} = 0.018...$$

To implement this, calculate the probability, then pick a random number between 0 and 1. If you pick a smaller number than your *transition probability*, let him go to the worse spot. This looks rather as though you are hitting the world with a hammer and jolting him about.

You can build on your hill climbing algorithm, considering a step left or right, along with a random move. You can also shrink the step sizes each time. Your algorithm looks like this:

```
pos = bag.left()
height = pos.y

while temperature > -5:
  if temperature < 0:
    step /= 2.0
  possible = [left(), right(), something_else()]
  for pos in possible:
    if pos.y < height or jump():
      height = pos.y
  temperature -= 0.1

return height
```

You now have all of the parts you need to get the turtle into the paper bag.
Time to code it.

Let's Find the Bottom of the Bag

Make a class to demonstrate a turtle moving around the paper bag:

Optimize/demo.py

```
Line 1  class Demo:
          def __init__(self, f):
            self.alex = turtle.Turtle()
            self.alex.shape("turtle")
     5      self.f = f

          def bag(self, points):
            line = turtle.Turtle()
            line.pen(pencolor='brown', pensize=5)
    10      line.up()
            line.goto(points[0], self.f(points[0]))
            line.down()
            for x in points:
              line.goto(x, self.f(x))
    15      line.hideturtle()

          def start(self, x):
            self.alex.hideturtle()
            self.alex.up()
    20      self.alex.goto(x, self.f(x))
            self.alex.down()
            self.alex.showturtle()
            self.alex.pen(pencolor='black', pensize=10)
            self.alex.speed(1)
    25
          def move(self, x, y, jump=False):
            if jump: self.alex.up()
            self.alex.goto(x, y)
            if jump: self.alex.down()
```

This stores a function (f), which describes the shape of the paper bag. Give the turtle a name, on line 4. Now you can have more than one turtle exploring your problem by making a Demo for each turtle you need. Your turtle, alex, will start where you tell him. You can get him to draw the bag, and then show him how to move to the points your optimizer picks. When he moves, he may jump as you see on line 27. Your hill climbing algorithm doesn't need this, but simulated annealing does.

Represent Paper Bags With Functions

Start with a slightly slanting bag. Write a function to tell you the y coordinate for an x coordinate:

```
Optimize/into_bag.py
def slanty_bag_curve(x):
  left = 0.5
  width = 9.
  if x < left:
    y = -20.*x+10.
  elif x < width + left:
    y = 0
  else:
    y = 20.*x-190
  return y
```

Your previous paper bags have been rectangular. For one x value you would then have several y values, so you cannot write this as a mathematical function. You can still make hill climbing work by coding this up with a generator to walk from left to right, or vice versa. Let's keep it simple here though.

For other functions, like the sloping cosine you saw earlier on page 190 use a lambda:

```
f = lambda x: 5*math.cos(x) - x
```

To make your turtle to walk into the bottom of the bag, you need an optimizer. You can try your optimizers against any function you can dream up. Let's start with hill walking.

Hill Climbing Algorithm

Make a hill_climb.py file, and add a single method named seek to this:

```
Optimize/hill_climb.py
Line 1  def seek(x, step, f):
  -       height = f(x)
  -       while True:
  -         if f(x-step) < height:
  5           x -= step
```

```
    elif f(x+step) <= height:
      x += step
    else:
      break
10  height = f(x)
    yield x, height
```

You have an initial x point, a fixed step, and a function (f). You can extend this to change the step size. In fact, you'll change the step size in simulated annealing, coming up soon. You'll see the difference this makes when you consider if your optimizers work later on. First things first.

Find the height at x. Consider a step left on line 4. If that goes down, great. Move left. If not, consider going right instead on line 6. Using less than or equal rather than strictly less than means alex continues right along a flat line, for example, the bottom of a bag. This isn't better, strictly speaking. It's up to you, but you'll get a chance to think about this later on. This simple hill climbing picks the first better spot. For problems with more than one dimension to search, you can try a few different directions and pick the best of these for a *steepest ascent/decent hill climbing* flavor. You can try *stochastic hill climbing* too by randomly picking a neighbor, making better neighbors more likely. You know how to use roulette wheel selection to ensure this. However you pick the points, once none are better, you stop generating new points. Your turtle can't step to a lower spot, so he stops.

To demonstrate the turtle walking into the bag, let's use the curve and demo like this:

Optimize/into_bag.py
```python
def slanty_bag():
  turtle.setworldcoordinates(-2.2, -2, 12.2, 22)
  demo = Demo(slanty_bag_curve)
  demo.bag([x*0.5 for x in range(-1, 22)])

  x = -0.5
  step = 0.1
  demo.start(x)
  gen = hill_climb.seek(x, step, slanty_bag_curve)
  for x, y in gen:
    demo.move(x, y, False)
```

Call setworldcoordinates to set the edges of the world. Set up your Demo, draw the bag, and then seek the best spot. Remember to start the turtle. This sets his position and speed. Call move while your hill_climb generates new points.

Simulated Annealing Algorithm

Alex the turtle gets to the bottom of the bag when he uses hill descent. Provided the bag is not crumpled up. If you try the slanting cosine, and he starts on the left, he gets to the first dip. Look back to the picture on page 190 for a reminder. If you use simulated annealing instead, he might reach the lower down dip.

Make a new file, sim_anneal.py. Like the hill climber, you try a left and right spot a step away. You'll also try a random point. If one is better than the current position, move there. If not, calculate a transition probability based on the current point, the new point, and a temperature:

Optimize/sim_anneal.py
```python
def transitionProbability(old_value, new_value, temperature):
    if temperature <= 0:
        return 0
    return math.exp((old_value - new_value) / temperature)
```

Remember, a zero probability means impossible. You stop jumping to worse spots when the temperature cools off. If you draw a random number smaller than the probability, you'll move to the worse spot.

Add a seek method:

Optimize/sim_anneal.py
```python
Line 1  def seek(x,
             step,
             f,
             temperature,
    5        min_x=float('-inf'),
             max_x=float('inf')):
          best_y = f(x)
          best_x = x
          while temperature > -5:
    10        jump = False
             if temperature < 0: step /= 2.0
             possible = [x - step, x + step, x + random.gauss(0, 1)]
             for new_x in [i for i in possible if min_x < i < max_x]:
                 y = f(new_x)
    15          if y < best_y:
                    x = new_x
                    best_x = new_x
                    best_y = y
                 elif transitionProbability(best_y, y, temperature) > random.random():
    20              jump = True
                    x = new_x
          yield best_x, best_y, temperature, jump
          temperature -= 0.1
```

You optionally send in the left and right edges (min_x and max_x) to stop your turtle wandering off the screen. You loop around until the temperature has dropped off. Once it's hit freezing, start shrinking the step sizes on line 11. This lets the turtle zone into the lowest point with more precision. You can go back and change the step size in hill climbing similarly.

This version stops when it gets very cold. However, you can keep going until the turtle stops moving instead, as you did for the hill descent. Find the possible points, a step to the left and right, and a random Gaussian step using Python's gauss method in the random library. This can give a large jump but will tend to give numbers close to zero, with a variance of 1, meaning large jumps are unlikely. Some implementations will scale this by the temperature, and you may need to experiment for harder problems. Harold Szu and Ralph Hartley discuss using occasional large jumps in their "Fast simulated annealing" 1984 paper.[1] If a possible point is better, go there. If not, find the transitionProbability and pick the worse place if you draw a smaller random number.

To demonstrate the turtle walking into the bag, it's worth setting up a utility function like this:

Optimize/into_bag.py
```
def sa_demo(curr_x,
        step,
        f,
        temperature,
        x_points,
        min_x, max_x,
        *setup):
    turtle.setworldcoordinates(*setup)
    demo = Demo(f)
    demo.start(curr_x)
    demo.bag(x_points)
    gen = sim_anneal.seek(curr_x, step, f, temperature, min_x, max_x)
    for x, y, t, j in gen:
        demo.move(x, y, j)
        curr_x = x
    print(curr_x, f(curr_x))
```

The setup controls the window size. The seek method gives you a generator so you can show the points the turtle explores. You set up the Demo as before. Get the x and y coordinates from the generator in a loop. To see if he jumped, look at j. The turtle doesn't draw a line when this happens. You can use the temperature (t) too to change the color of the line so you can watch how your system cools. You can print out the final spot visited if you wish.

1. www.researchgate.net/publication/234014515_Fast_simulated_annealing

Use the sa_demo to see if the turtle finds the lowest point in the slopey cosine:

Optimize/into_bag.py
```
def sa_cosine_slope(bounded):
  f = lambda x: -x+5*math.cos(x)
  x_points = [x*0.1 for x in range(-62, 62)]
  min_x, max_x = bounds(bounded, x_points)
  temperature = 12
  step = 0.2
  sa_demo(x_points[0], step, f, temperature,
          x_points,
          min_x, max_x,
          -6.2, -12, 6.2, 12)
```

If you don't set limits on the turtle's exploration, he can wander off the edge of the screen. You can set bounds to keep the turtle between the left and right of the bag, or give him free reign:

Optimize/into_bag.py
```
def bounds(bounded, x_points):
  if bounded:
    return x_points[0], x_points[-1]
  return float('-inf'), float('inf')
```

Try varying the starting point, initial step size, and temperature. Try him out on your other paper bag too. Try other curves. What happens if you use a cosine instead? All you need to do is use a different function:

```
f = lambda x: 10*math.cos(x)
```

Scaling up by 10 makes the curve larger, so it's easier to see.

With one turtle, you're only going to find one "best" point. Try two turtles at random starting points. Try several. Can you find both of the lower down points?

Did It Work?

Whichever function and algorithm you chose, you might end up with a turtle in your bag. If you don't set the limits in the simulated annealing, you might end up with a turtle off the edge of the screen.

Hill Climbing

With the slightly slanting bag, and a sensible step size, around 0.1, your turtle can settle down in the bottom of the bag as shown in the figure on page 199.

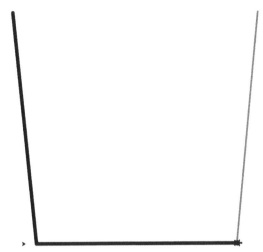

He's settled on the far right corner of the bag. If you make a larger step size, for example, the whole width of the bag, the turtle will get stuck quickly. You see this clearly when you try a different function, such as

```
f = lambda x: math.fabs(x)
```

This gives you a v-shape. If you can't imagine this as a paper bag, think of it as a cross-section through a paper cone. The lowest point is at (0, 0). Can the turtle get there?

Make a demo to find out:

Optimize/into_bag.py
```
Line 1  def stuck():
            turtle.setworldcoordinates(-12, -1, 12, 15)
            f = lambda x: math.fabs(x)
            demo = Demo(f)
       5    start = -10
            step = 3
            demo.start(start)
            demo.bag(range(-10, 11))
            gen = hill_climb.seek(start, step, f)
       10   for x, y in gen:
                demo.move(x, y, False)
```

When you start at -10, a step of 10 on line 6 is fine, taking the turtle straight to the lowest point. A step of 3 goes right a bit but stops before the lowest point. The turtle tries (-10, 10), (-7, 7), (-4, 4), (-1, 1) An extra step would be back uphill to (2, 2), so the turtle stops. If you start a little further along, at x=-9 you can get to the lowest point. The starting position makes a difference for this algorithm. A step of 20 from (-10, 10) goes right and stops there. At (10, 10), a step 20 right is uphill, and 20 left is back to the start, so has the original height. The next picture shows these step size of 3, then 10 then 20:

The step size affects where your turtle stops. He goes left if that's better (using strictly less) or right if the point is at least as good (using less than or equal). Your turtle, therefore, goes right as far as possible on a flat line. If you use less than or equal for both directions, the turtle would ping back and forth. In the official lingo, the algorithm does not *converge*. It's often worth deciding a maximum number of iterations to avoid pinging backward and forward between spots forever. The tendency to go right in this algorithm avoids the problem. You have used a set number of epochs for many algorithms in this book. They ensure you stop looping, even if you don't get convergence.

The paper cone has one lower spot. The slanting cosine has two dips on the screen, one lower than the other. Your turtle finds one when he hill climbs, using small steps. He does not find the lowest spot though:

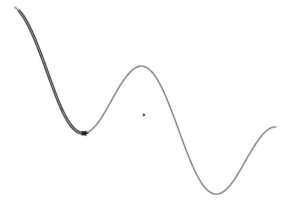

By changing the step size as the algorithm runs, you can do better with the fabs function. However, your hill climbing will always stop in the first dip encountered. The turtle gets stuck in a local minimum. He cannot find the better spot in the cosine curve. This is why you also tried simulated annealing.

Simulated Annealing

Your turtle can get to the bottom of the first slanting bag:

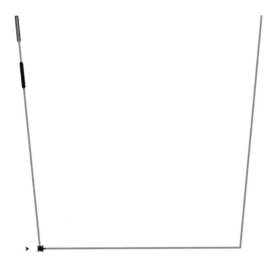

He doesn't leave a black line all the way along—you can see the jumps taken. You'll get slightly different jumps each time you run this. Whichever path the turtle takes, he ends up at the bottom of the bag. These jumps help him find the lowest point. Make sure you use the bounded version of the sa_demo. Otherwise, the determined turtle may go to a farther down spot off the screen. When bounded, he picks the lower, right dip:

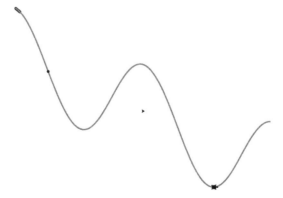

Again, the turtle has jumped several times, so the black path taken isn't continuous. He has found the best spot though. A grand improvement on hill climbing.

Faced with a cosine curve, your solitary turtle will have to pick one spot. He cannot be in two places at once. Try having three turtles.

Optimize/into_bag.py
```
def sa_cosine_turtles(bounded):
  turtle.setworldcoordinates(-6.2, -12, 6.2, 12)
  curr_x = [-6.0, 0, +6.0]
  f = lambda x: 10*math.cos(x)
  count = 3
  demo = [Demo(f) for _ in range(count)]
  x_points = [x*0.1 for x in range(-62, 62)]
  demo[0].bag(x_points)
  min_x, max_x = bounds(bounded, x_points)
  gens = []
  temperature = 10.0
  step = 0.2
  for i, x in enumerate(curr_x):
    demo[i].start(curr_x[i])
    gens.append(sim_anneal.seek(x, step, f, temperature, min_x, max_x))
  for (x1, y1, t1, j1), (x2, y2, t2, j2), (x3, y3, t3, j3) in zip(*gens):
      demo[0].move(x1, y1, j1)
      demo[1].move(x2, y2, j2)
      demo[2].move(x3, y3, j3)
```

If they start in three different places—left, middle, and right—you almost always end up with a turtle or two in each dip:

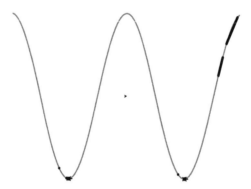

In fact, this even works if you pick random starting points for each using random.choice(x_points).

Your turtles can find their way into the paper bags, provided you try a few random spots or get lucky with your step size and keep them on the screen.

Extension to More Dimensions

So far, all of the paper bags were two-dimensional. Since the turtle moves along the bag edge, defined by a function like y=f(x), this is a one-dimensional optimization problem. You can extend hill climbing to paper bags of any shape in three, or more dimensions, thereby exploring higher dimensional optimization problems. Instead of a choice between left or right, you then have other additional directions to consider. You can also try *gradient descent* methods when you have more dimensions. Hill climbing steps in one direction along an axis. Gradient descent combines directions, to get something like northeast, by calculating the gradient and taking the steepest slope down or up. You need to do some mathematics to calculate the best gradient step.

When you hill climb or descend you consider two steps f(x-step) and f(x+step). You compare these to the current value f(x) so you can approximate the gradient using

$$\frac{f(x-step)-f(x)}{-step}$$

$$\frac{f(x)-f(x+step)}{step}$$

The bigger negative gradient takes you downhill. The bigger positive gradient takes you uphill. Put a turtle at (-2, 4) on a quadratic curve

```
f = lambda x: x**2
```

If he takes a step of three right or left, he's approximating the gradient:

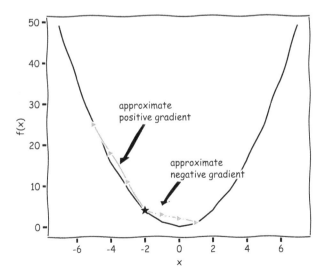

As you shrink the step size, you get closer to the precise value of the gradient or *derivative* of the function. Your turtle can then move to the bottom of the bag quickly. He can use the technique to explore three-dimensional paper bags.

For a multi-dimensional problem, your function

$$f(\overline{x}) = f(x_1, x_2, ..., x_n)$$

has a gradient called ∇f. This is a vector made up of the partial derivatives or slopes in each direction.

$$\nabla f = \left(\frac{\partial f}{\partial x_1}, \frac{\partial f}{\partial x_2}, ..., \frac{\partial f}{\partial x_n} \right)$$

Find the biggest gradient, either using the approximation with division or working through the math, and take a step in that direction. In two dimensions, this might be straight north or a linear combination like north-east. The step size is usually called the *learning rate*, γ. You move from a point

$$\overline{p} = (p_1, p_2, ..., p_n)$$

to the next using

$$\overline{p_{n+1}} = \overline{p_n} - \gamma \nabla f(\overline{p_n})$$

This steps you toward the minimum of f. Loop around until you see no improvement or start zig-zagging. However complicated the algorithm, the idea is still similar to your turtle climbing down into the bag. Find a good step and walk the line until you're done.

Imagine your paper cone in three dimensions. If you look at it from above, and sketch lines at the same height, you see circles:

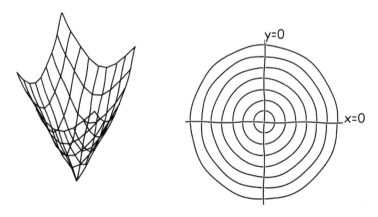

A hill climbing turtle, shown as a solid line in the next picture, steps in an axis direction. A gradient descent turtle can take a more direct route, shown as a dotted line:

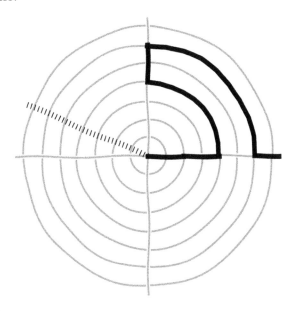

Over to You

You used simple hill climbing to get your turtle to the bottom of the bag. He moves left or right, provided such a point is at least as good as the current spot. He gets to the bottom of the bag, provided you use the right step size, get lucky with the starting point, and you don't crumple up the bag too much. You then moved on to simulated annealing, allowing you to hit things with a (virtual) hammer until they work. The algorithms are similar, both considering a left or right step. However, SA will try a random, perhaps worse point, once in a while. As with many of the algorithms you've seen in this book, trying something random can help you solve problems. As the systems cool, random jumps become less likely.

You can try various other cooling schemes. A common method decreases the temperature by a multiple each time, somewhere between 0.8 and 0.99. The method you use can make a huge difference. It also depends on how you choose neighbors or potential spots.[2] For the simple problem in this chapter, it doesn't make much difference. You can alter the temperature and random steps considered together to build an *adaptive simulated annealing* algorithm. The variant in this chapter shrunk the step size once the temperature dropped

---

2. https://en.wikipedia.org/wiki/Simulated_annealing#Cooling_schedule

below zero—at that point, no more jumps happened. Instead, try a wide potential jump initially, and shrink this as the temperature cools.

By using three turtles, you touch upon the idea of a *niching method*. You can adapt your genetic algorithms or swarm algorithms to have more than one population. This gives you more than one overall best solution and can help you solve problems with several good spots. There are several variants of niching, some allowing *fitness sharing*.[3] You have one population but ensure you include different solutions by adapting your selection procedure.

Many other machine learning algorithms—such as neural networks—need gradient descent methods to find the best model. *Stochastic gradient descent* (SGD) can be used to find a line dividing several points. This uses training data and draws a line between items in one class and those in another. You encountered other classifiers in Chapter 2, *Decide! Find the Paper Bag*, on page 15. SGD shuffles the training data and iteratively moves the line until you find the best boundary. Optimization is used in many machine learning algorithms. Go try out some you haven't covered yet. However complicated the algorithms look, imagine your turtle considering how to get nearer his goal.

You've seen a variety of ways to escape a paper bag using many different machine learning algorithms. There are many, many more you can learn about, and the list will continue to grow. Of course, some will need some serious mathematics. You can, and probably should, use a framework to do the hard work for you. Numerical computing is difficult. However, you now have a feel for how many algorithms work. Start somewhere, possibly random. Loop around, tweaking variables as you go. Stop when you've found something good enough. Test what you've done. And, most importantly, try changing your parameters to see what happens. Over to you.

---

3. https://stackoverflow.com/questions/37836751/what-s-the-meaning-of-fitness-sharing-and-niche-count-in-ga

Bibliography

[DS04]   Marco Dorigo and Thomas Stützle. *Ant Colony Optimization*. MIT Press, Cambridge, MA, 2004.

[Hul06]   John C. Hull. *Options, Futures and Other Derivatives*. Prentice Hall, Englewood Cliffs, NJ, 2006.

[MP69]   Marvin Minsky and Seymour Papert. *Perceptrons: an introduction to computational geometry*. MIT Press, Cambridge, MA, 1969.

[Pet08]   Charles Petzold. *The Annotated Turing: A Guided Tour through Alan Turing's Historic Paper on Computability and the Turing Machine*. John Wiley & Sons, New York, NY, 2008.

[Tor15]   Adam Tornhill. *Your Code as a Crime Scene*. The Pragmatic Bookshelf, Raleigh, NC, 2015.

Index

Thank you!

How did you enjoy this book? Please let us know. Take a moment and email us at support@pragprog.com with your feedback. Tell us your story and you could win free ebooks. Please use the subject line "Book Feedback."

Ready for your next great Pragmatic Bookshelf book? Come on over to https://pragprog.com and use the coupon code BUYANOTHER2019 to save 30% on your next ebook.

Void where prohibited, restricted, or otherwise unwelcome. Do not use ebooks near water. If rash persists, see a doctor. Doesn't apply to *The Pragmatic Programmer* ebook because it's older than the Pragmatic Bookshelf itself. Side effects may include increased knowledge and skill, increased marketability, and deep satisfaction. Increase dosage regularly.

And thank you for your continued support,

Andy Hunt, Publisher

Fix Your Hidden Problems

From technical debt to deployment in the very real, very messy world, we've got the tools you need to fix the hidden problems before they become disasters.

Software Design X-Rays

Are you working on a codebase where cost overruns, death marches, and heroic fights with legacy code monsters are the norm? Battle these adversaries with novel ways to identify and prioritize technical debt, based on behavioral data from how developers work with code. And that's just for starters. Because good code involves social design, as well as technical design, you can find surprising dependencies between people and code to resolve coordination bottlenecks among teams. Best of all, the techniques build on behavioral data that you already have: your version-control system. Join the fight for better code!

Adam Tornhill
(274 pages) ISBN: 9781680502725. $45.95
https://pragprog.com/book/atevol

Release It! Second Edition

A single dramatic software failure can cost a company millions of dollars—but can be avoided with simple changes to design and architecture. This new edition of the best-selling industry standard shows you how to create systems that run longer, with fewer failures, and recover better when bad things happen. New coverage includes DevOps, microservices, and cloud-native architecture. Stability antipatterns have grown to include systemic problems in large-scale systems. This is a must-have pragmatic guide to engineering for production systems.

Michael Nygard
(376 pages) ISBN: 9781680502398. $47.95
https://pragprog.com/book/mnee2

Learn Why, Then Learn How

Get started on your Elixir journey today.

Adopting Elixir

Adoption is more than programming. Elixir is an exciting new language, but to successfully get your application from start to finish, you're going to need to know more than just the language. You need the case studies and strategies in this book. Learn the best practices for the whole life of your application, from design and team-building, to managing stakeholders, to deployment and monitoring. Go beyond the syntax and the tools to learn the techniques you need to develop your Elixir application from concept to production.

Ben Marx, José Valim, Bruce Tate
(242 pages) ISBN: 9781680502527. $42.95
https://pragprog.com/book/tvmelixir

Programming Elixir 1.6

This book is *the* introduction to Elixir for experienced programmers, completely updated for Elixir 1.6 and beyond. Explore functional programming without the academic overtones (tell me about monads just one more time). Create concurrent applications, but get them right without all the locking and consistency headaches. Meet Elixir, a modern, functional, concurrent language built on the rock-solid Erlang VM. Elixir's pragmatic syntax and built-in support for metaprogramming will make you productive and keep you interested for the long haul. Maybe the time is right for the Next Big Thing. Maybe it's Elixir.

Dave Thomas
(410 pages) ISBN: 9781680502992. $47.95
https://pragprog.com/book/elixir16

Books on Python

For data science and basic science, for you and anyone else on your team.

Data Science Essentials in Python

Go from messy, unstructured artifacts stored in SQL and NoSQL databases to a neat, well-organized dataset with this quick reference for the busy data scientist. Understand text mining, machine learning, and network analysis; process numeric data with the NumPy and Pandas modules; describe and analyze data using statistical and network-theoretical methods; and see actual examples of data analysis at work. This one-stop solution covers the essential data science you need in Python.

Dmitry Zinoviev
(224 pages) ISBN: 9781680501841. $29
https://pragprog.com/book/dzpyds

Practical Programming, Third Edition

Classroom-tested by tens of thousands of students, this new edition of the best-selling intro to programming book is for anyone who wants to understand computer science. Learn about design, algorithms, testing, and debugging. Discover the fundamentals of programming with Python 3.6—a language that's used in millions of devices. Write programs to solve real-world problems, and come away with everything you need to produce quality code. This edition has been updated to use the new language features in Python 3.6.

Paul Gries, Jennifer Campbell, Jason Montojo
(410 pages) ISBN: 9781680502688. $49.95
https://pragprog.com/book/gwpy3

A Better Web with Phoenix and Elm

Elixir and Phoenix on the server side with Elm on the front end gets you the best of both worlds in both worlds!

Programming Phoenix 1.4

Don't accept the compromise between fast and beautiful: you can have it all. Phoenix creator Chris McCord, Elixir creator José Valim, and award-winning author Bruce Tate walk you through building an application that's fast and reliable. At every step, you'll learn from the Phoenix creators not just what to do, but why. Packed with insider insights and completely updated for Phoenix 1.4, this definitive guide will be your constant companion in your journey from Phoenix novice to expert, as you build the next generation of web applications.

Chris McCord, Bruce Tate and José Valim
(325 pages) ISBN: 9781680502268. $45.95
https://pragprog.com/book/phoenix14

Programming Elm

Elm brings the safety and stability of functional programing to front-end development, making it one of the most popular new languages. Elm's functional nature and static typing means that run-time errors are nearly impossible, and it compiles to JavaScript for easy web deployment. This book helps you take advantage of this new language in your web site development. Learn how the Elm Architecture will help you create fast applications. Discover how to integrate Elm with JavaScript so you can update legacy applications. See how Elm tooling makes deployment quicker and easier.

Jeremy Fairbank
(250 pages) ISBN: 9781680502855. $40.95
https://pragprog.com/book/jfelm

JavaScript and more JavaScript

JavaScript is back and better than ever. Rediscover the latest features and best practices for this ubiquitous language.

Rediscovering JavaScript

JavaScript is no longer to be feared or loathed—the world's most popular and ubiquitous language has evolved into a respectable language. Whether you're writing frontend applications or server-side code, the phenomenal features from ES6 and beyond—like the rest operator, generators, destructuring, object literals, arrow functions, modern classes, promises, async, and metaprogramming capabilities—will get you excited and eager to program with JavaScript. You've found the right book to get started quickly and dive deep into the essence of modern JavaScript. Learn practical tips to apply the elegant parts of the language and the gotchas to avoid.

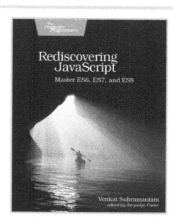

Venkat Subramaniam
(286 pages) ISBN: 9781680505467. $45.95
https://pragprog.com/book/ves6

Simplifying JavaScript

The best modern JavaScript is simple, readable, and predictable. Learn to write modern JavaScript not by memorizing a list of new syntax, but with practical examples of how syntax changes can make code more expressive. Starting from variable declarations that communicate intention clearly, see how modern principles can improve all parts of code. Incorporate ideas with curried functions, array methods, classes, and more to create code that does more with less while yielding fewer bugs.

Joe Morgan
(282 pages) ISBN: 9781680502886. $47.95
https://pragprog.com/book/es6tips

More on Java

Get up to date on the latest Java 8 features, and take an in-depth look at concurrency options.

Functional Programming in Java

Get ready to program in a whole new way. *Functional Programming in Java* will help you quickly get on top of the new, essential Java 8 language features and the functional style that will change and improve your code. This short, targeted book will help you make the paradigm shift from the old imperative way to a less error-prone, more elegant, and concise coding style that's also a breeze to parallelize. You'll explore the syntax and semantics of lambda expressions, method and constructor references, and functional interfaces. You'll design and write applications better using the new standards in Java 8 and the JDK.

Venkat Subramaniam
(196 pages) ISBN: 9781937785468. $33
https://pragprog.com/book/vsjava8

Programming Concurrency on the JVM

Stop dreading concurrency hassles and start reaping the pure power of modern multicore hardware. Learn how to avoid shared mutable state and how to write safe, elegant, explicit synchronization-free programs in Java or other JVM languages including Clojure, JRuby, Groovy, or Scala.

Venkat Subramaniam
(280 pages) ISBN: 9781934356760. $35
https://pragprog.com/book/vspcon

The Pragmatic Bookshelf

The Pragmatic Bookshelf features books written by developers for developers. The titles continue the well-known Pragmatic Programmer style and continue to garner awards and rave reviews. As development gets more and more difficult, the Pragmatic Programmers will be there with more titles and products to help you stay on top of your game.

Visit Us Online

This Book's Home Page
https://pragprog.com/book/fbmach
Source code from this book, errata, and other resources. Come give us feedback, too!

Keep Up to Date
https://pragprog.com
Join our announcement mailing list (low volume) or follow us on twitter @pragprog for new titles, sales, coupons, hot tips, and more.

New and Noteworthy
https://pragprog.com/news
Check out the latest pragmatic developments, new titles and other offerings.

Save on the eBook

Save on the eBook versions of this title. Owning the paper version of this book entitles you to purchase the electronic versions at a terrific discount.

PDFs are great for carrying around on your laptop—they are hyperlinked, have color, and are fully searchable. Most titles are also available for the iPhone and iPod touch, Amazon Kindle, and other popular e-book readers.

Buy now at *https://pragprog.com/coupon*

Contact Us

Online Orders:	*https://pragprog.com/catalog*
Customer Service:	*support@pragprog.com*
International Rights:	*translations@pragprog.com*
Academic Use:	*academic@pragprog.com*
Write for Us:	*http://write-for-us.pragprog.com*
Or Call:	+1 800-699-7764

Milton Keynes UK
Ingram Content Group UK Ltd.
UKHW010155260924
448856UK00007B/205